Welcome to…

iPhone for Beginners

Time and time again, the iPhone has redefined the idea of what a phone can be. Apple's most iconic device has come a long way since the debut of its original model in 2007, but that hasn't stopped the company from seeking to push the boundaries of mobile technology with every new release. With the iPhone 15 range, as well as the introduction of iOS 17, Apple has once again set the standard for smartphones in quality, versatility and accessibility. One of the only downsides to having such a feature-packed phone is that it can be difficult to know where to start. Thankfully, this new edition of iPhone for Beginners covers everything you need to know. From setting up your new iPhone to getting the most from the pre-installed apps, editing your photos to getting directions, Apple Music to Apple Pay – it's all here. Get to grips with all the best features via our step-by-step tutorials and see how your device can transform your user experience. Enjoy the book!

FUTURE

iPhone for Beginners

Future PLC Quay House, The Ambury, Bath, BA1 1UA

Editorial
Compiled by **Stephen Williams**
Features Editor **Aiden Dalby**
Senior Art Editor **Andy Downes**
Head of Art & Design **Greg Whitaker**
Editorial Director **Jon White**

Photography
James Sheppard
All copyrights and trademarks are recognised and respected

Advertising
Media packs are available on request
Commercial Director **Clare Dove**

International
Head of Print Licensing **Rachel Shaw**
licensing@futurenet.com
www.futurecontenthub.com

Circulation
Head of Newstrade **Tim Mathers**

Production
Head of Production **Mark Constance**
Production Project Manager **Matthew Eglinton**
Advertising Production Manager **Joanne Crosby**
Digital Editions Controller **Jason Hudson**
Production Managers **Keely Miller, Nola Cokely, Vivienne Calvert, Fran Twentyman**

Printed in the UK

Distributed by Marketforce, 5 Churchill Place, Canary Wharf, London, E14 5HU
www.marketforce.co.uk Tel: 0203 787 9001

iPhone for Beginners Twenty-Sixth Edition (TCB5711)

FUTURE

Connectors. Creators. Experience Makers.

Future plc is a public company quoted on the London Stock Exchange (symbol: FUTR)
www.futureplc.com

Chief executive **Zillah Byng-Thorne**
Non-executive chairman **Richard Huntingford**
Chief Financial and Strategy Officer **Penny Ladkin-Brand**

Tel +44 (0)1225 442 244

Part of the iCreate bookazine series

Contents

8 Essential guide to iPhone
Get to know your iPhone and see how it can enrich your life both at home and at work

APP OF THE DAY

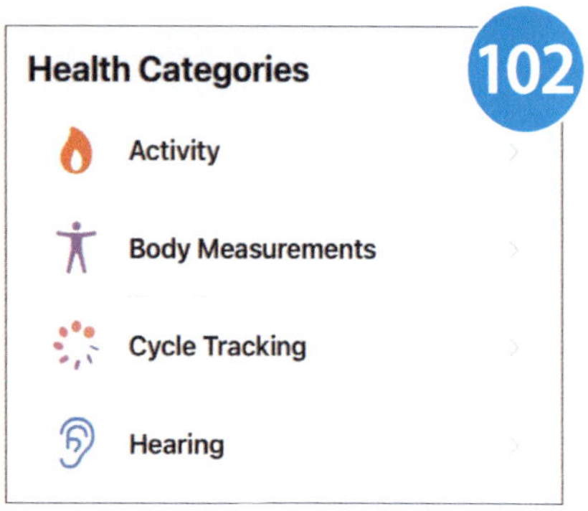

Health Categories
102
Activity
Body Measurements
Cycle Tracking
Hearing

"You can do even more with apps from the App Store"

"Master the basics and then move on to the next level – there will be no stopping you!"

The next step

112

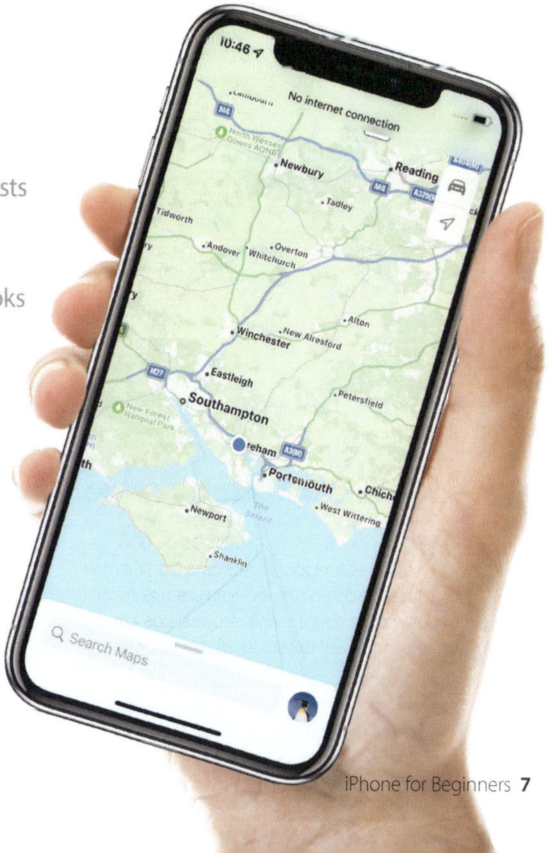

The essential iPhone guide

Everything you need to know to navigate and use your phone like a pro

Image: Getty Images

Now that you have finally got your hands on your new iPhone, you'll no doubt be keen to master it. In truth, it couldn't be easier. The operating system (currently iOS 17) is intuitive and simple to use, and all of the key apps come pre-installed, visible on your Home screen. Before you get to experience these though you will need to go through an initial setup process when firing up your device for the first time.

This takes the form of several steps that explain the features that need to be set up and guide you through the process of activating them. The first step is creating an Apple ID (or signing in if you already have one). This is linked to all aspects of your iPhone, from buying apps and items from the iTunes Store to utilising Apple's excellent iCloud service – all that is required is a valid email address and a secure password.

A lot of the services and features that are set up during this process are optional and can be tweaked later if preferred. For this the Settings app is your best friend; all of the different facets of your iPhone are presented in a simple list, so just tap on a category and begin exploring the many options associated with it.

Once you are set up, take time to explore the many core apps that come pre-installed on your device, such as Safari (for web browsing), Mail (for checking your emails), Calendar (your personal organiser) and News (for up-to-the-minute reports). Before too long you'll be chomping at the bit to download new apps and games from the App Store. An array of options are available, some of which require payment, while others are totally free – such as Facebook and Instagram. If you're keen to start listening to music you can go to the iTunes Store or Apple Music app.

Many of these features are the product of years of development and a major new iOS is released every year (and regularly updated) to introduce new features that make using your iPhone to manage your life much easier.

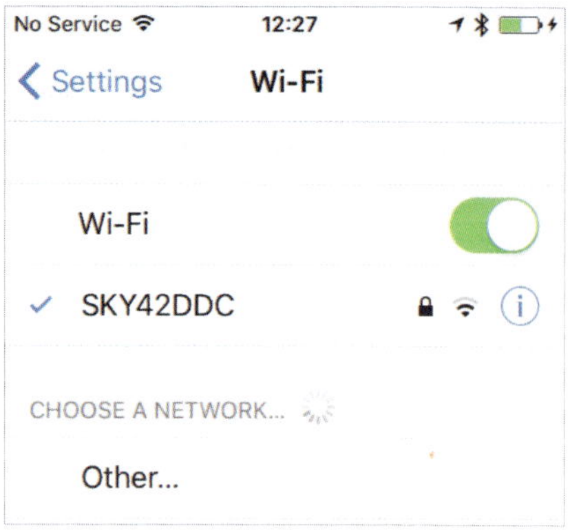

Fig: To get started, connect to a Wi-Fi network and start shopping for new apps and content

"Once you are set up, take time to explore the many core apps that come pre-installed on your device, such as Safari, Calendar and News. Before too long you'll be chomping at the bit to download new apps and games from the App Store"

Explore the iPhone SE's hardware

Learn how to find your way around your device

Navigate your iPhone
Your iPhone is a wondrous device packed into a compact little shell. Here we explain what all of the external buttons do

Volume buttons
The volume buttons on the side of your device are used to control sound output. You can also assign functionality to the Camera app

Carrier
Your phone carrier will be displayed here, when you're connected to its network. The row of dots represents the strength of the signal you receive

Wi-Fi icon
If you are connected to a Wi-Fi network then the signal strength will be displayed by this icon

Dock connector
The port on the bottom of your iPhone is used by the lightning cable to connect your device to a power source or computer

Battery icon
This icon shows the current battery level of your device. Go to Settings>General> Usage to view it as a percentage figure

Power button
This button on the side of your device can be tapped to put your device to sleep or a long-press will power the phone down completely

Camera
Your iPhone comes with front- and rear-facing cameras. Look here and say 'cheese' for your perfect selfie!

Apps
The main home screen displays most of the pre-installed apps in a grid system. Tap on an app to launch it

Home button
Pressing this button can wake your device up and close apps to return to the main Home screen

First steps

Take the time to learn the basics of your iPhone, which we highlight here

Turn on/off

How to power-up your device or shut it down completely using the power button

01 To turn on your iPhone, hold down the power button at the right until the logo is displayed

02 If your iPhone is in sleep mode, press the power button once to wake it up or press the Home button

03 To turn your iPhone off, hold the power button until the screen dims and move the slider right

Sleep mode

Conserve your iPhone's battery power when the device is not in use

01 Go to Settings>Display & Brightness and then tap Auto-Lock to set how soon your iPhone will automatically lock and go to sleep when idle

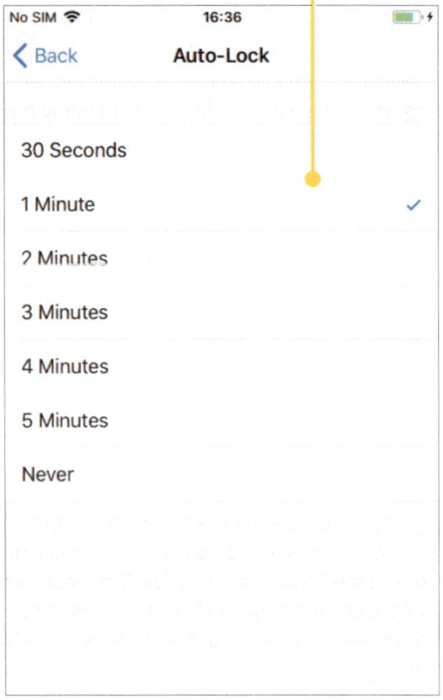

Unlock

Determine how your iPhone is unlocked ready for use

01 Go to Face ID & Passcode and turn Passcode on. When setting a code, if you tap 'Passcode Options' you can set a more complex code

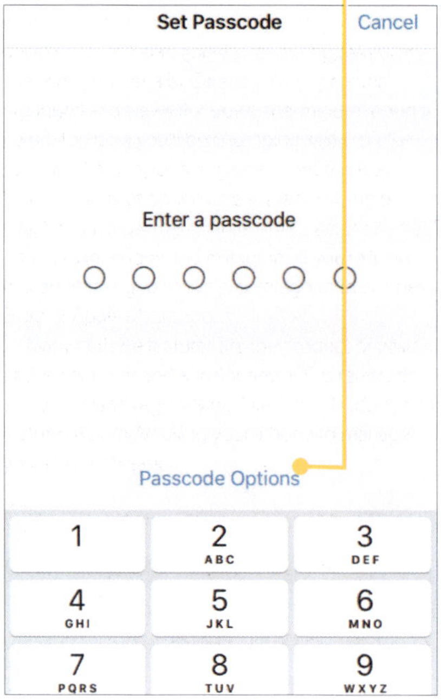

Charging

All you need to know about powering-up your iPhone

01 To charge your iPhone, use the lightning cable to connect to a plug socket or computer and it will be indicated by the icon next to the battery

Change volume

Discover how to adjust the sound output of your iPhone for all kinds of purposes

01 On your home screen, use the volume buttons on the side of the iPhone to adjust the ringer volume. Do the same in an app to adjust the volume.

02 You can also swipe up from the bottom of the screen to access the Control Centre and then swipe left to see the volume controls

Mute

Learn how to cut all sound on your device instantly

01 To mute the ringer instantly, flick the switch above the volume buttons on your iPhone. Repeat to unmute.

Rotation lock

Learn how to lock your iPhone's screen orientation to keep it in landscape or portrait mode

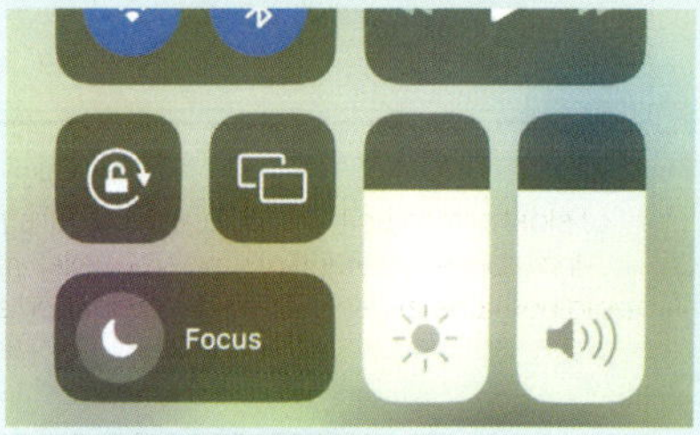

01 Position your device in the orientation you wish to lock it in and then swipe up from the bottom of the screen to access your Control Centre.

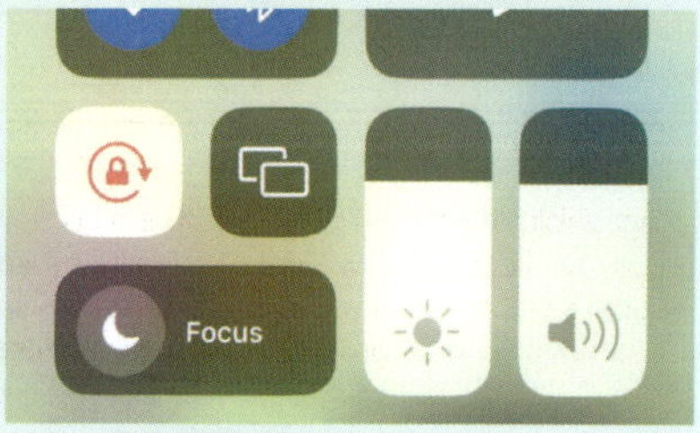

02 Now tap the top-left icon (with the padlock on) to lock the current orientation. Repeat the process to unlock it again.

Syncing

iCloud ensures all of your info and files stay up to date

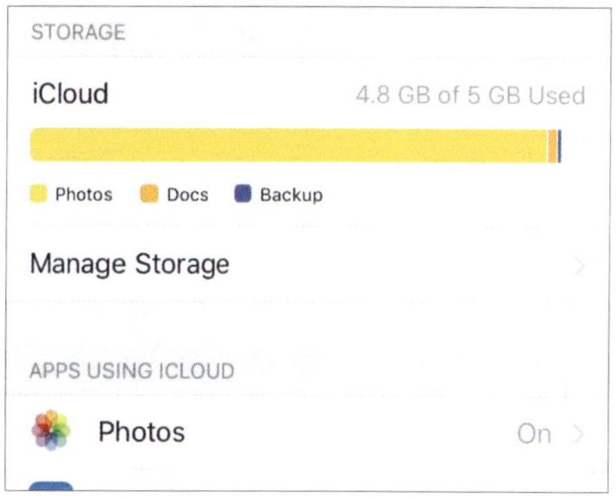

01 Sign in using your Apple ID and then move the sliders to enable the service for various apps.

Brightness

Learn how to adjust the brightness of your iPhone screen to suit any conditions

01 Adjusting screen brightness can be done by going to the Settings app and tapping on Display & Brightness or swiping up to access Control Centre.

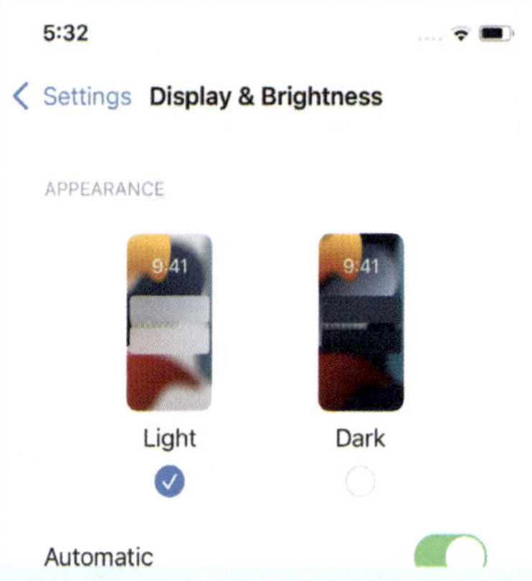

Icons

How to organise your apps neatly on your Home screen

01 Apps can be moved by pressing and holding on an icon, pressing Edit Home Screen and then dragging the icon into a new position.

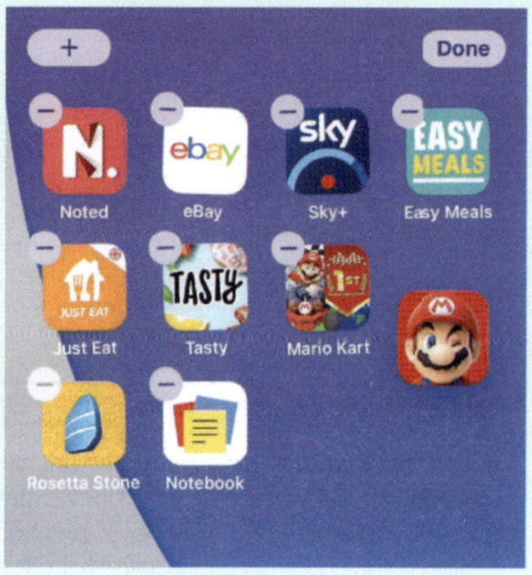

Folders

Folders are a great way to help keep apps organised

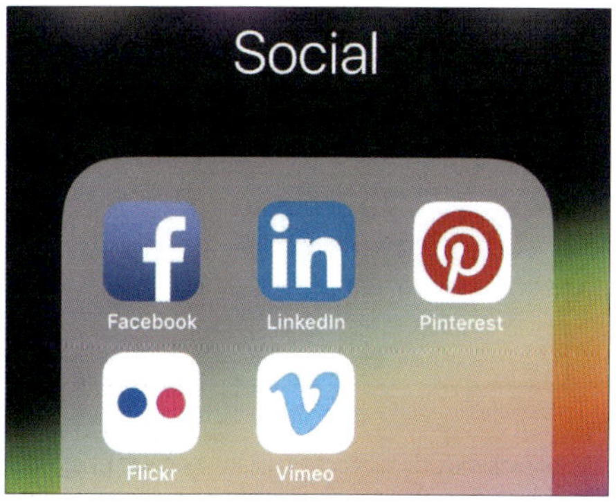

01 Press and hold on an app icon until they all shake and then drag an icon on top of another icon to create a folder.

Safari

Safari is your iPhone's default web browsing app and it comes packed with features to make web trawling easier. The URL field at the top of the app doubles up as a search field (just type in keywords to perform searches) and the 'aA' icon to the left of the field initiates Reader mode, which presents web pages in a stripped-down, easier-to-read format. The icons at the bottom of the interface allow you to view your open tabs and share web content via social media or into your Notes. The book icon provides access to your bookmarks, items added to your Reading List for offline reading and your browsing history. Using the tabs icon you can also view and close open tabs, open a private tab and scroll down to see web pages open on other iOS and macOS devices that you can access. In the Settings app you can changed the default search engine in Safari.

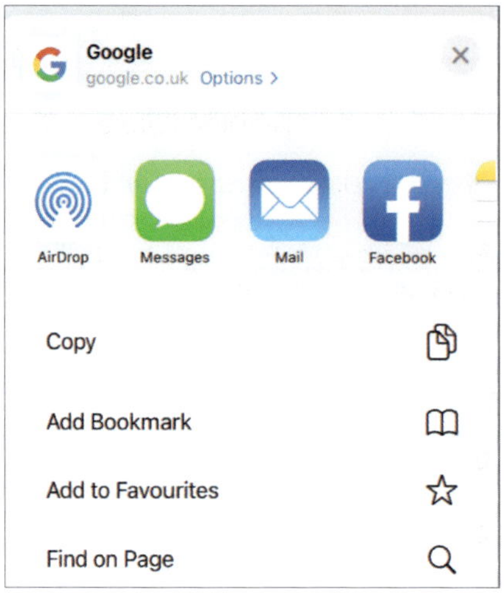

01 Tapping the share icon while on a web page throws up numerous options to share content across multiple platforms. You can also add pages to your Reading List.

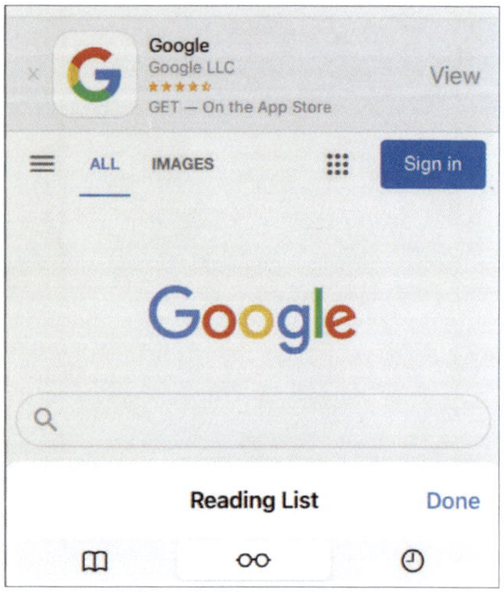

02 The book icon lets you access your bookmarks, Reading List (which lets you read saved pages offline) and also view your internet history.

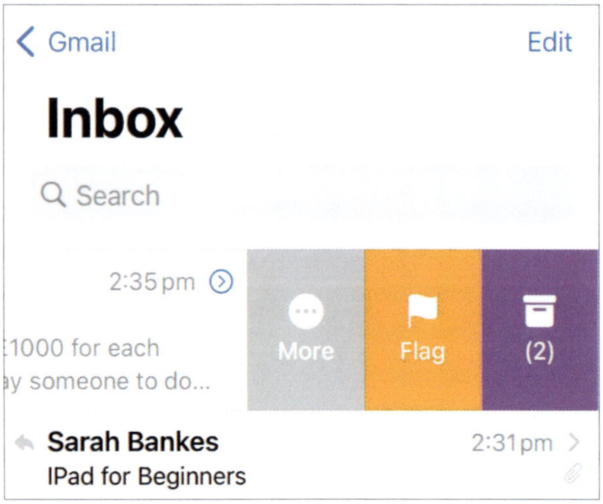

Mail

For all of your emailing needs, the Mail app is your iPhone's default email app and is feature-rich for all of your email management needs. When launching the Mail app for the first time you will need to go through an initial set-up process – but this involves little more than entering a valid email address and password. Once done, you can start sending emails almost instantly by tapping on the pen and paper icon in the lower-right corner of the interface. Simply enter the recipient(s), add a subject and then enter your body text. If you wish to send media attachments then you can do this by tapping the share icon while in other apps (such as Photos) and then choosing the email option. All of your emails can be organised into neat little folders to make viewing and managing them easier and you can also mark people as VIPs so that their mails go into a special priority inbox. While in your mailbox, swipe left on an email to see options to Flag, Archive and more.

App Store

The App Store is your first port of call for purchasing and downloading new iPhone apps. When you launch the app for the first time you will be sent to the Today page, which flags up collections of the most exciting content as well as providing an app and game of the day. There are often also 'Meet the Developer' features, which are interviews with the people behind some of the store's most popular apps. Read them to get an insight into their work, and then you can follow the link to their app or game at the bottom of the screen. To start exploring in more detail, there are icons at the bottom of the interface for Apps, Games, Search, and Arcade. Both Games and Apps have a top seller list on each of their pages so you can find the most popular downloads. Tap on the profile icon at the top to see all of the possible updates for the apps that you own (that you can download for free), and you can also view apps you have purchased in the past and re-download them to your current device.

Tuesday 10 Calendar

Stay organised in your professional and personal life

Views
You can change the calendar view by tapping in the top-left corner. Switch from days, to months and years. Weeks can be scrolled in day view

+ (add event)
You can create new calendar events by tapping once on the '+' icon (which stands for 'add event'). Here you can add titles, times, locations and even more

Calendars
You can view all of your calendars and enable/disable them from displaying in the app

Agenda
Tap on the list icon at the top of the interface to see your agenda. Scroll down through the various days to see what events are lined up

Reminders

Get acquainted with the app that never lets you forget…

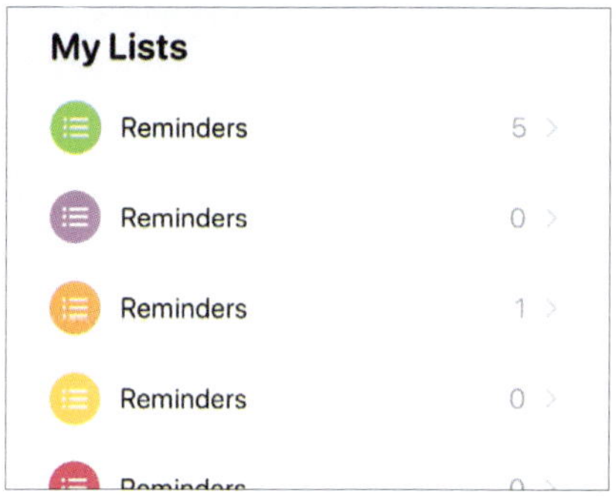

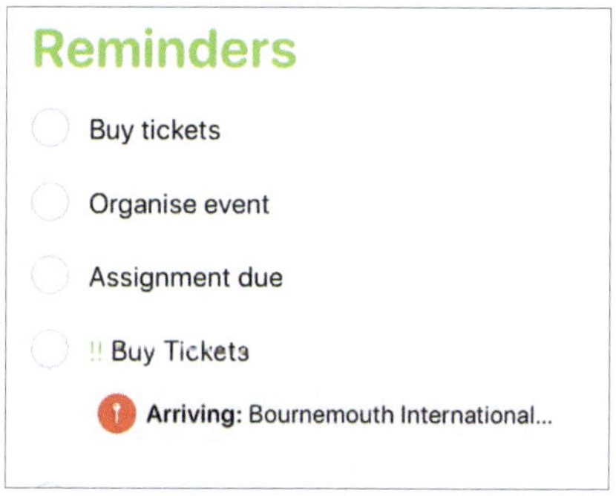

01 You can create a new list by tapping on the '+' icon. Give your new list a name to make it easy to track and then colour-code it for further easy reference. When you have finished, tap 'Done' to return to the lists screen.

02 Tap on a line in a list to add a new reminder item. Tap the small 'i' icon next to an entry to add more details, so as a time to be reminded and a location that triggers a reminder when you are in range.

Read a book on your iPhone

Books

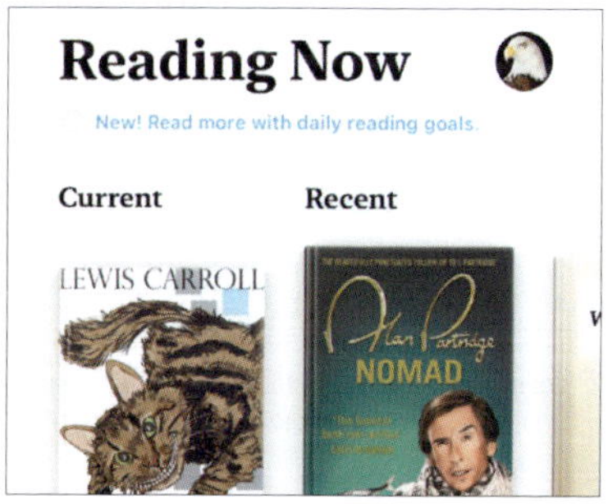

The Books app now comes pre-installed on your device. From within this app you can use the Book Store, Audiobooks and Search icons at the bottom of the interface to browse the store for fresh content and see all of your books displayed on shelves. Tap to open it and see a range of options to tailor the experience and add notes and bookmarks.

Monitor your fitness levels

Health

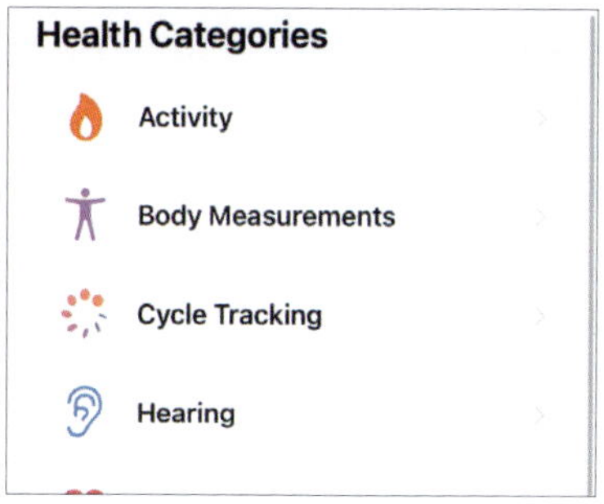

The Health app allows you to link all of your favourite fitness and diet apps and get all of the data streamed into one place to help you easily monitor your fitness levels. You can also create a personal medical ID card in the app that can be accessed from your Lock screen in the case of an emergency.

Maps

Explore the world with your iPhone's built-in navigation system

Search for locations
Enter place names, addresses or post codes into the search field to find locations. You can then get directions to and from those locations across an array of different transportation methods

'i' (options)
Tap this icon to change the style of the map displayed , display live traffic and drop a pin in your current location

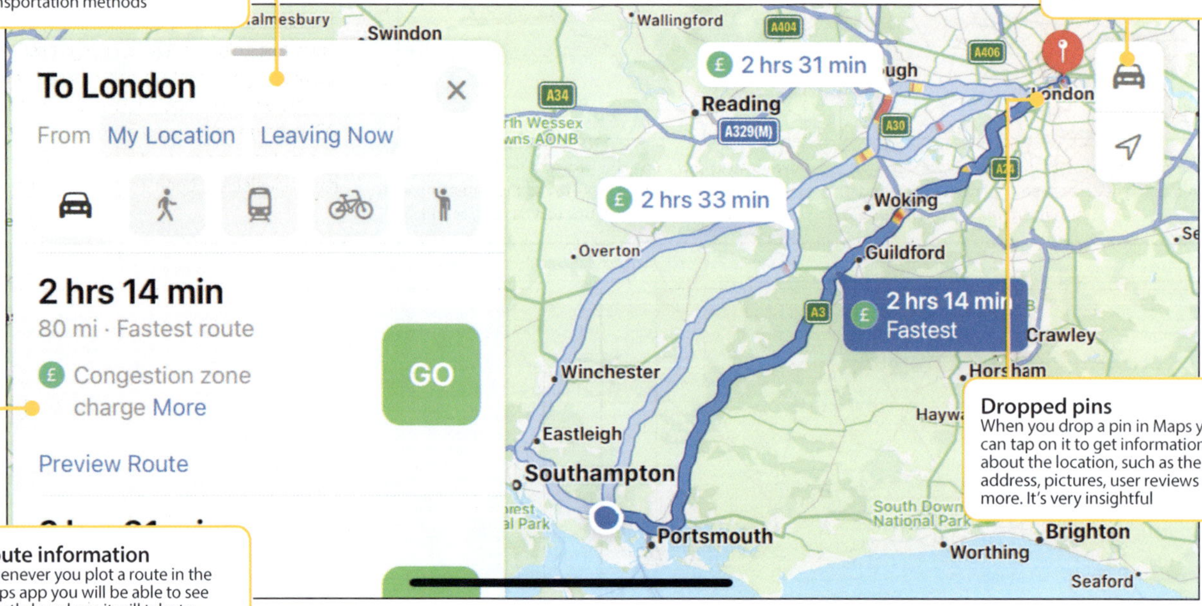

Dropped pins
When you drop a pin in Maps you can tap on it to get information about the location, such as the address, pictures, user reviews and more. It's very insightful

Route information
Whenever you plot a route in the Maps app you will be able to see exactly how long it will take to reach your destination and whether any toll roads are included

Camera & Photos

Your iPhone's Camera and Photos apps have been intrinsically linked to make capturing and editing images as simple as possible. The Camera app features a dynamic set of shooting modes and live filters (as well as features such as a shutter delay timer) and the Photos app groups your captured images together into time frames and locations. You can also perform a surprisingly versatile sequence of edits and share them with friends and family. You can jump straight to your iPhone's Photos app from within the Camera app by tapping on the thumbnail of the last image you took and the Camera app itself can be launched via your Home screen, Control Centre or by swiping up on the camera icon on the Lock screen to ensure that you never miss a moment.

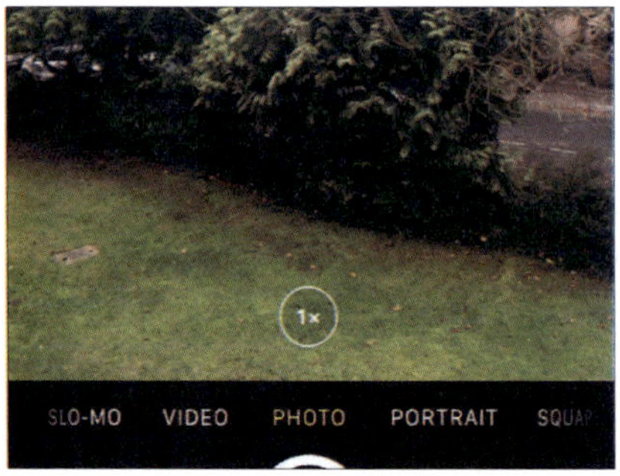

01 While in your iPhone's Camera app, swipe through the various modes displayed just above the shutter button. You can also apply live filter effects to your photos by tapping on the filter icon in the upper-right corner. There is a whole range of filters and effects to choose from. You will also discover the options to enable or disable the iPhone's flash, enable HDR (high dynamic range) mode and set a timer at the top.

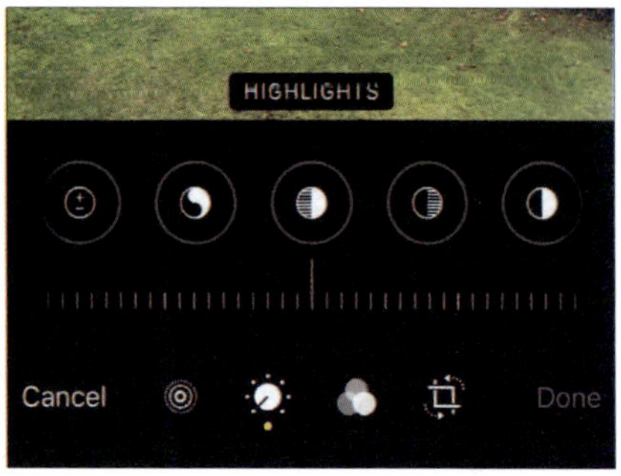

02 The Photos app on your iPhone lets you view your captured images, share them with other people, mark them as favourites, delete them or perform some incredible edits. Tapping on the Edit option provides options to crop your image, change the filter, adjust the colour levels and more. You can also auto-enhance the images to save you time. Once you have applied your edits, just tap Done.

News

Get the latest news headlines delivered straight to your device

Your personal edition
The News app brings all of the latest news from your favourite sources and delivers it all in a tidy, dated edition that you can scroll through and read at your leisure

Favourite sources
The news that is presented is taken from sources that you hand-pick. This means that you are only getting the news that you want to read without having to wade through the stuff you don't

Instant reaction
As you scroll through the categories you can swipe an article then tap on the thumbs up icon and this will inform the News app of what stories interest you

News+
For a monthly fee you can have access to hundreds of magazines and newspapers from around the world as well as exclusive content. You can preview the content here

iTunes & Music

The iTunes store, accessible from its own app on your iPhone's Home screen, is a sprawling online shop for music and visual entertainment. You can buy or rent movies or TV shows to view in your Apple TV app or shop and download tracks or albums to listen to in your Music app. New content is well flagged up and the comprehensive search engine means that you're never far away from what you want to watch or listen to. Any music you buy will be automatically downloaded to your Music app and here you will be able to create playlists. All of your purchased music is also synced to iCloud Drive. This means that if you put your iPhone down and switch to your iPad, you can open the iTunes app and tap the cloud icon to re-download it onto your current device.

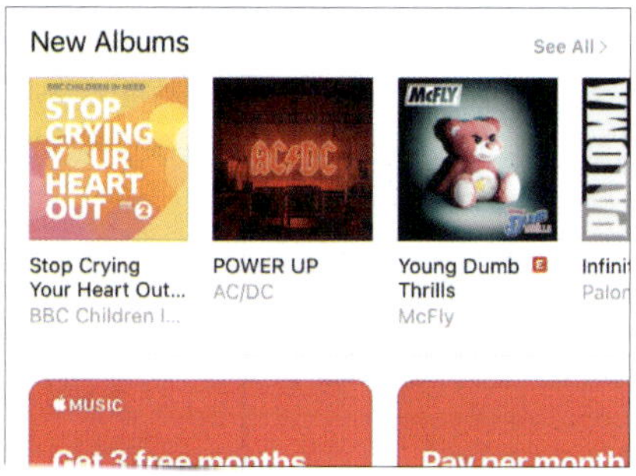

01 The iTunes app is where you go to shop for new music, movies, and TV shows to download and enjoy on your iPhone (or any other iDevice you may own, such as an iPad). iTunes is a constantly changing environment as new content arrives on the iTunes store to purchase all the time. Use the icons at the bottom of the app interface to jump to the various sections and begin browsing content – including the Search tab, which you can tap on to see what is currently trending.

02 All of the music that you purchase and download through the iTunes app will be made playable through your iPhone's Music app. Tap to open up the Music app and view your downloaded music. Then just tap on individual tracks to begin playing them. You can create your own original playlists to suit any occasion or mood – or just let the Genius function do it for you. Genius Mixes are preselected compilations of songs from your music library.

Explore the iPhone's hardware

Learn how to find your way around Apple's revolutionary device

New design
The iPhone 15 and iPhone 15 Pro feature an Super Retina XDR display. The cases have a ceramic shield with aluminium edges so they're lightweight while still being able to be dropped without breaking

Face ID
The front-facing camera can scan the user's face to confirm their identity. This can be used to unlock the phone or confirm payments using Apple Pay

TrueDepth camera
With an infrared camera, ambient light sensor, proximity sensor and flood illuminator the iPhone X can take selfies better than any other iPhone

Large Power button
Now called the 'Side button,' not only did it see an increase in size, it can also be used to activate Siri by long-pressing or press it twice to bring up Apple Pay

Super Retina display
The iPhone 15 range use XDR displays. The 6.1 screen has a resolution of 2556x1179 at 460ppi, while the iPhone 15 Pro uses a 6.7 inch display with a resolution of 2796x1290 at 460ppi

Wireless charging
The newer iPhones are compatible with Qi-standard wireless charging technology. There are plenty of charging mats on the market already

No Home button
The Home button has been removed to make way for the all-screen design, but its functions have been moved to other buttons or changed to swiping gestures

Accessories

You can also connect other devices to your iPhone, such as keyboards and speakers. Here we explore some of the possibilities

Fitness armband
Keep your iPhone secure and close at hand to monitor your fitness and supply motivational music during your workouts with this attractive armband from Belkin

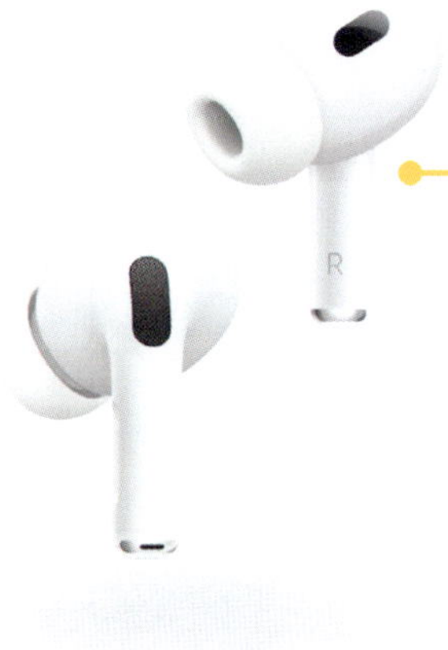

AirPods
These wireless earphones provide exceptional sound quality and include sensors that can detect when you are listening and when you are not so they can stop the music

Apple Watch
Get more from your iPhone with the Apple Watch. Accept calls, track your heartbeat, make payments and so much more with this brilliantly intuitive piece of kit

Case
Why not help preserve the wellbeing of your treasured iPhone by investing in a silicone case? These stylish covers from Apple come in a wide range of colours and offer plenty of spongy protection for your device

Guitar interface
You can plug your electric guitar straight into your iPhone using devices such as the IK Multimedia iRig HD-2 Guitar Interface and apply a range of cool effects to your sounds

Setting up

The first steps to take with your iPhone

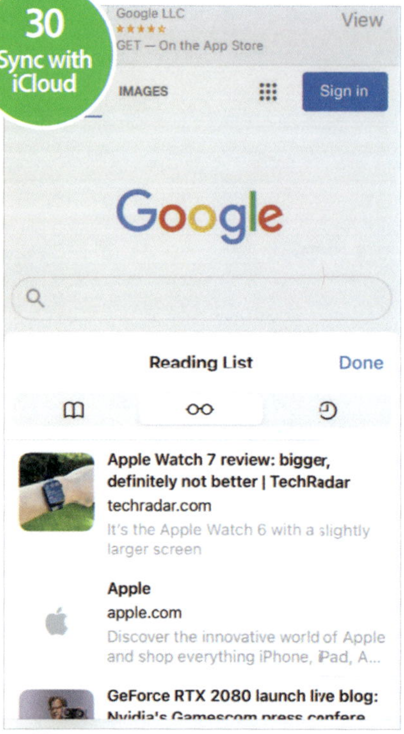

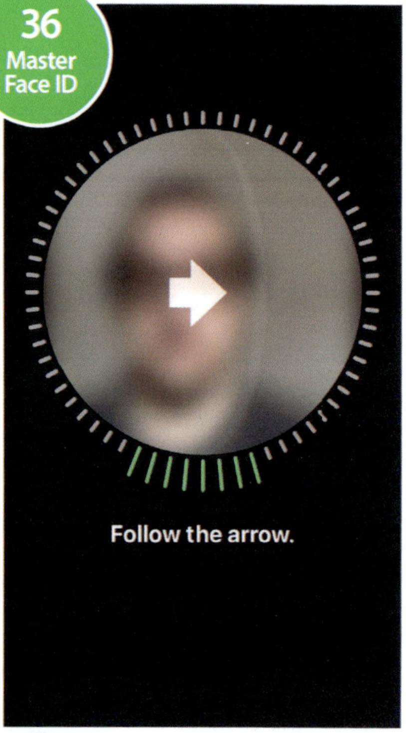

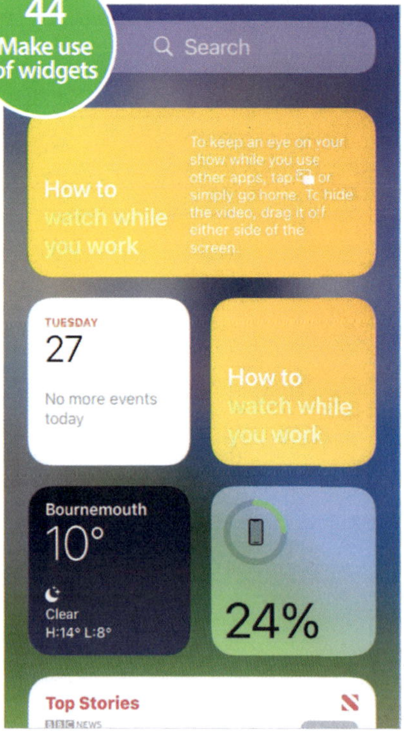

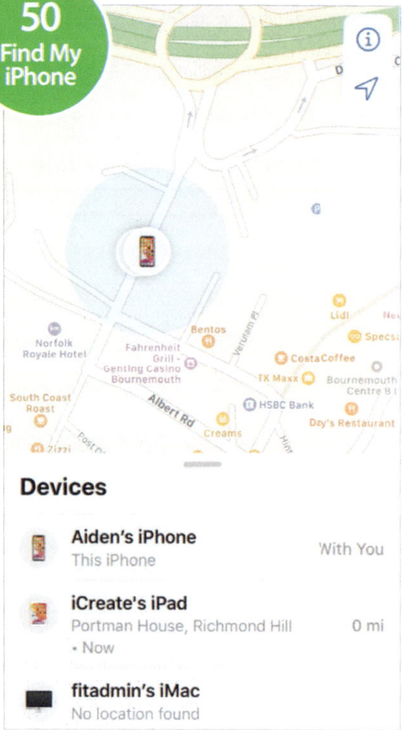

46
Customise Notification Centre
Notification Centre
While in Do Not Disturb
TIME SENSITIVE
Meeting
Traffic is light. It will take 57 min to get to Bournemouth International Centre Exeter Road, Bournemouth, BH2 5BH, England on M27.
in 22m
Leave Time Sensitive notifications from Calendar on? This allows Calendar to deliver important notifications immediately.
Leave On
Turn Off
iPhone Not Backed Up
40m ago
This iPhone hasn't been backed up in 114 weeks. Backups happen when this iPhone is connected to power, locked and on Wi-Fi.

Activate and register a new iPhone

Just got an iPhone? We guide you through the process of setting up your device and registering it to your Apple ID

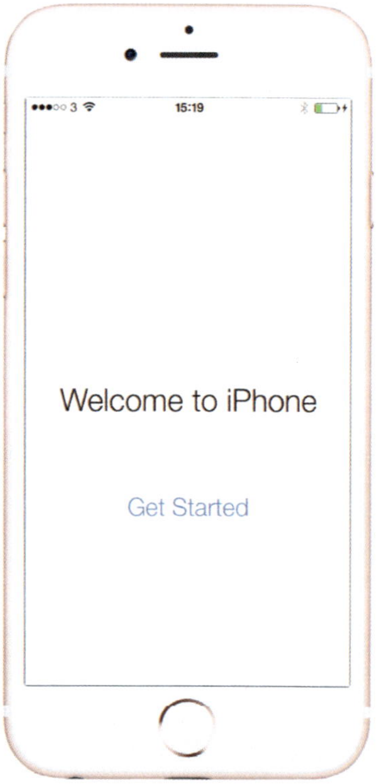

With trembling hands and a giddy sensation in your stomach, you unpack your brand new iPhone from its box and are ready to activate it and start using it to enhance your life. The entire setup can be carried out independently from your computer, which makes it a quick and easy process.

All you have to do is connect your new device to a power source and then press the Power button, which is situated on the top of the right-hand edge of your iPhone. It is on the top-right of the device if you have an older generation of iPhone. This will bring up a generic iPhone lock screen with a slider at the bottom. Tap and hold on the slider with your index finger and swipe it to the right to unlock the device and the next series of screens will guide you through the setup process.

In this tutorial we will take you through each stage of the process and explain what each screen is asking you to do, whether it is setting up your Apple ID, if you don't already have one from making purchases on iTunes, or connection to Wi-Fi.

Activation Prepare your iPhone

01 Unlock your device

When you first switch on your iPhone you will be presented with a plain-looking lock screen with a slider at the bottom. Tap and hold the slider and swipe it to the right to unlock the iPhone.

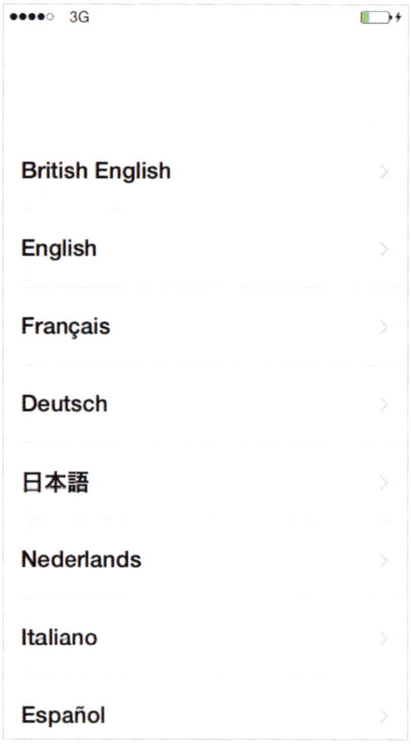

02 Select your language

You will be presented with a long list of different languages. The language you choose here will be the language that your device is set to and the language that all text will be presented in.

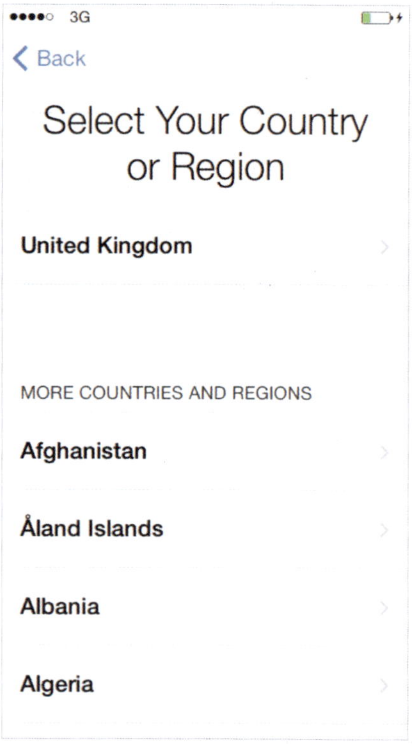

03 Select your Country or Region

Now select your Country or Region. This will determine the apps that are available in the App Store and the various media available from the iTunes app, as well as the correct currency.

Explore your Home screen

Once your iPhone has been set up, start exploring

Your Settings
Your first port of call once you leave the Dock should be the Settings app. Tap it to start modifying and customising aspects of your device

Your Dock
The Dock is a row of app icons in a strip at the bottom of the screen. This Dock is present no matter which of your screens you're on and should be used for your most-used apps

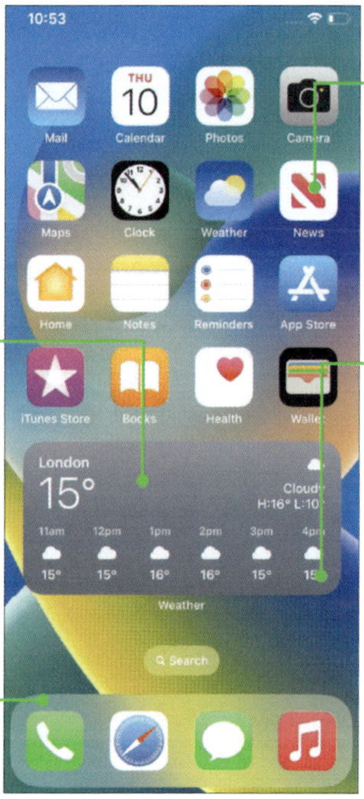

Your apps
A selection of Apple apps are built into the operating system and will appear on your Home screen as standard. Tap on an icon to launch the app

Adding apps
The Dock contains four apps as standard but you can have fewer or swap them out by tapping and holding on an icon until it starts to shake and then dragging it with your finger into the Dock

Tweaking options
If you couldn't decide whether to activate certain services such as Find My iPhone and iCloud during the initial setup process then you can do it later. Just tap on your Settings app and you will find the relevant options there to activate and start using these features.

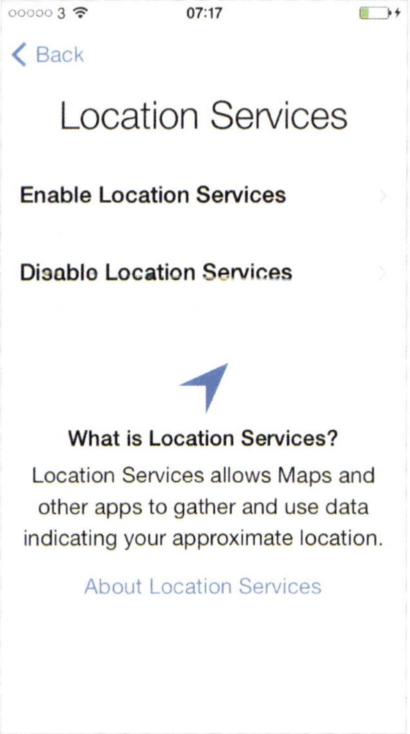

04 Location Services
The next part of the process is dedicated to your device's Location Services. This allows apps to gather and use data indicating your approximate location. This is useful if you lose your device.

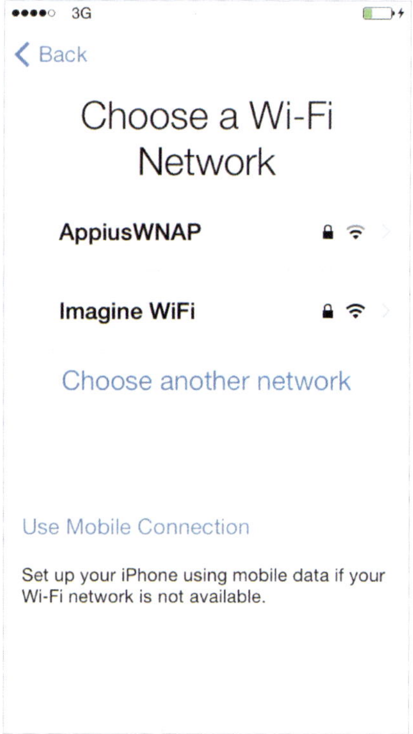

05 Connect to a network
Your iPhone relies on a Wi-Fi network to be able to connect to the internet and fuel a wealth of different services. If you are within range of a Wi-Fi network then it will be detected.

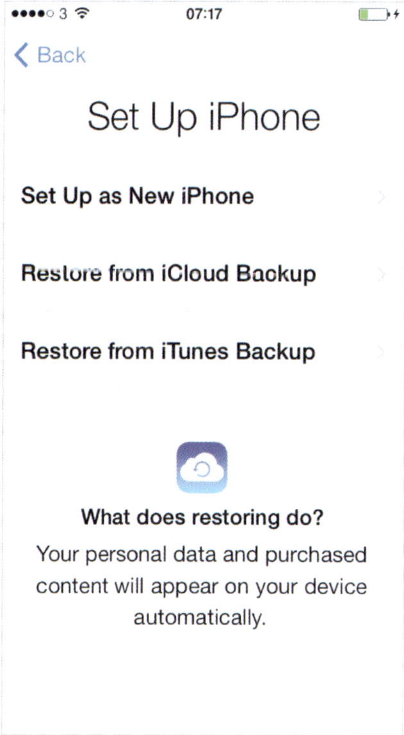

06 Set up your iPhone
You're given three options in the next step. You can set up the device as a new iPhone or restore it from previous settings backed up to either iCloud or iTunes.

Setting up

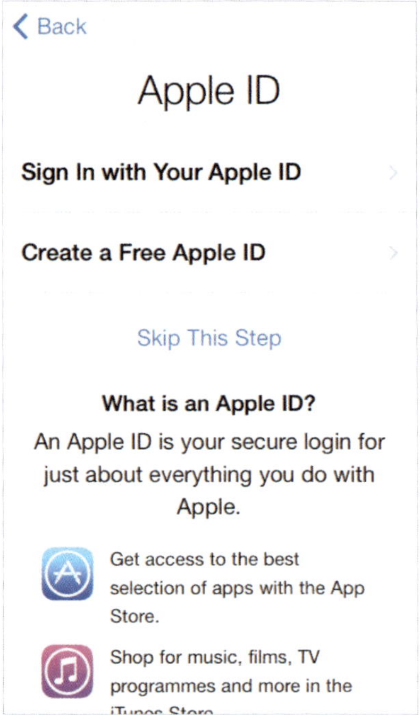

07 Sign in

You'll need an Apple ID to enjoy the main features of your iPhone, such as downloading apps and back up and sync with iCloud.

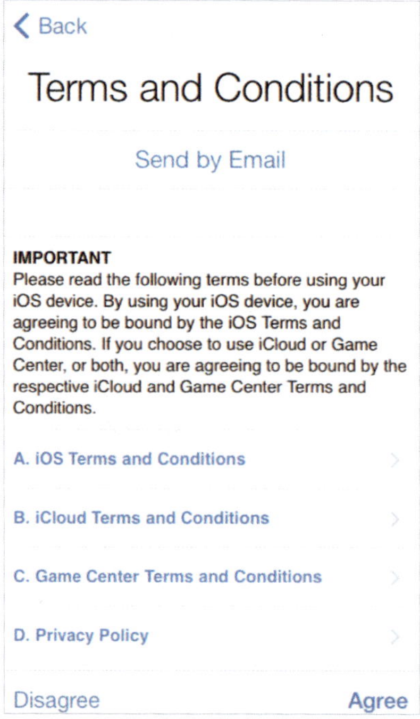

08 Read the T&Cs

After signing in, you will be presented with a screen of Terms and Conditions that relate to all aspects of your iPhone such as iCloud etc.

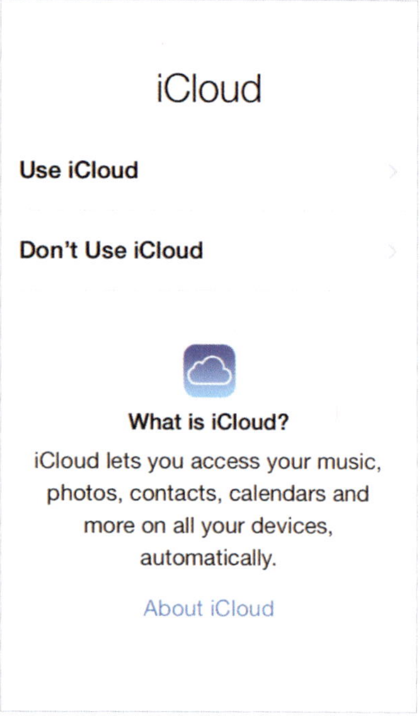

09 Set up iCloud

Apple's iCloud first came with iOS 5 and it allows you to back up data from your iPhone to your own free cloud storage space.

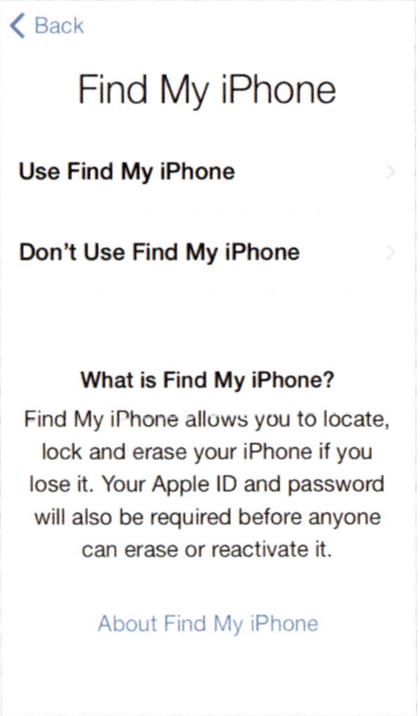

10 Find My iPhone

If you lose your iPhone, Find My iPhone will help you locate it. Activate this service to sync the location of your device with your iCloud.

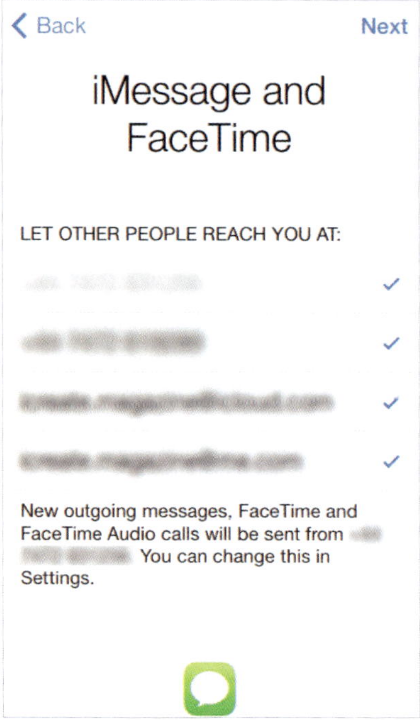

11 iMessage and FaceTime

FaceTime lets people speak face-to-face, while iMessage provides free messaging. Enter your email address or phone number to use them.

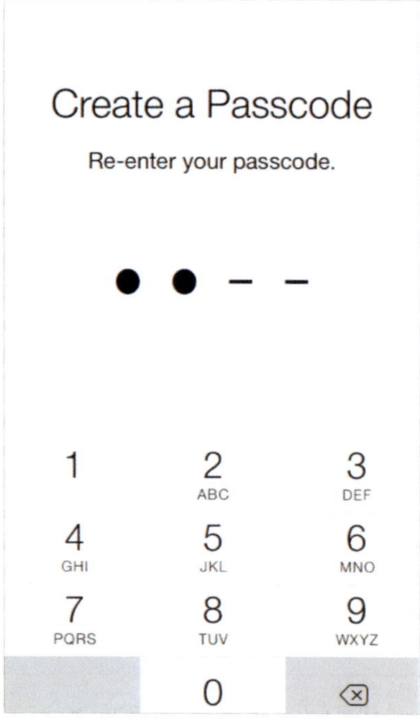

12 Create a passcode

Enter a four-digit passcode to secure your iPhone and prevent others using it. You'll need to enter the code every time you unlock your device.

Exploring your Settings

Start personalising your iPhone through Settings

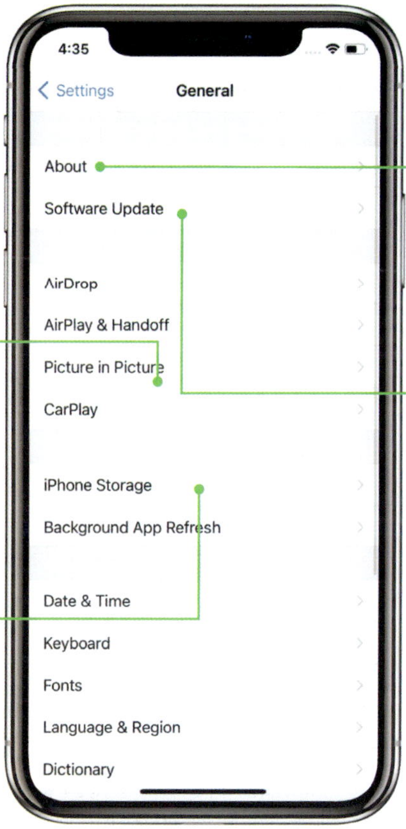

The categories
All of the categories to which the Settings apply will be presented in a list when you launch Settings. Tap an option to bring up its individual screen of settings

iPhone Storage
Here is where you can check to see how much storage you have left on your device, it's broken down by app, so you can see which are the space hoggers, as well as your iCloud storage and battery life

All about your iPhone
By tapping on the About option in the General section, you can get in-depth info on your device, such as the serial number, capacity and the amount of space that your various apps and media take up

Software Update
By tapping on the Software Update option in the General section your device will be able to find and update your iOS, without the need to connect your device to a computer

Restoring from backup
You may not be setting up a brand new iPhone; you may just be restoring your existing iPhone using an iCloud backup. In which case the setup process is still carried out the same way, but when you're asked to set up your iPhone, choose Restore from iCloud Backup and all of your previous settings will be restored and all of your apps and media will be re-downloaded. This can take a while.

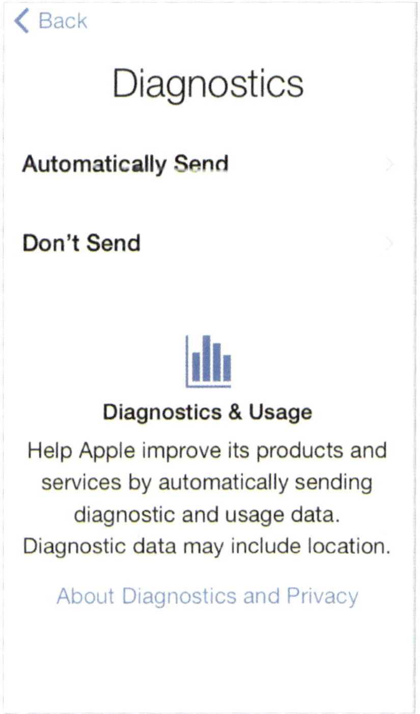

13 Diagnostics & Usage
Apple likes to keep track of how its products are performing, so this screen allows you to send diagnostic data straight to Apple.

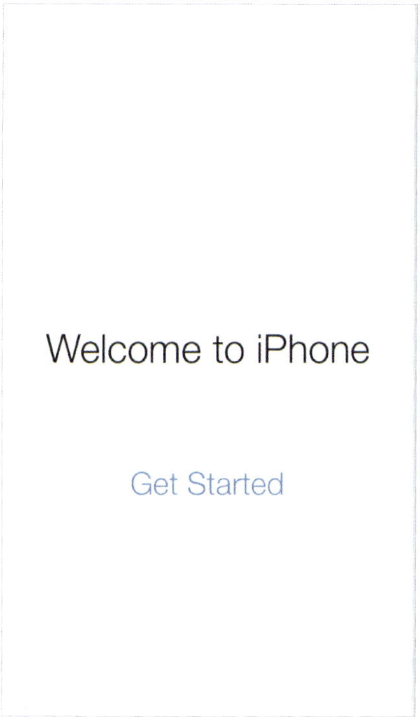

14 Set up complete
Congratulations! You have now worked your way through the entire set-up process. A screen will confirm the process is complete.

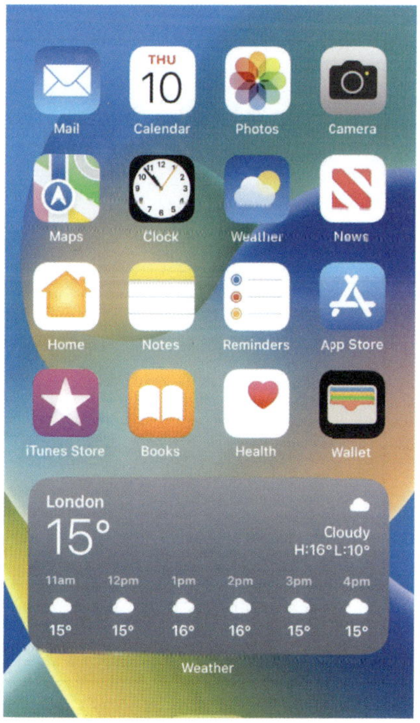

15 Start using your iPhone
You'll be presented with your Home screen, so start tapping on icons to launch apps and find your way around.

Get to grips with Settings

Understanding how to personalise your device can make all the difference. We show you how to make important choices

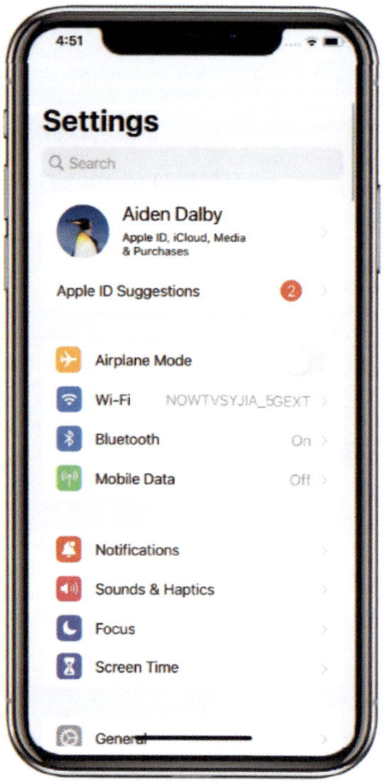

Your iPhone's operating system is highly configurable, and we don't just mean that you can change your wallpaper, though of course you can! The default settings when you first fire it up are pretty good for most people, but there will surely be some things that you would like it to do differently, whether it's waiting a little longer before locking the screen, remembering your passwords in Safari or adding a signature to your outgoing Mail messages. More or less all of the iOS apps like Mail, Messages, Music, Photos and the Phone section of an iPhone can be tweaked and tuned to behave as you like, whether it's using shake to shuffle tunes in your Music app or automatically checking for mail.

You can't alter every single thing of course, and though some people claim that this is a bad thing, quite the opposite is true. On mobile phones and tablets that do let you endlessly muck about with every part of the system, the result is almost always a terrible, unusable mess. Whereas the iPhone gives you just enough room to personalise your device while making sure that you don't end up actually making things worse.

> "The default settings are good, but there will be some things that you would like it to do differently"

Settings The basics of the iPhone's Settings app

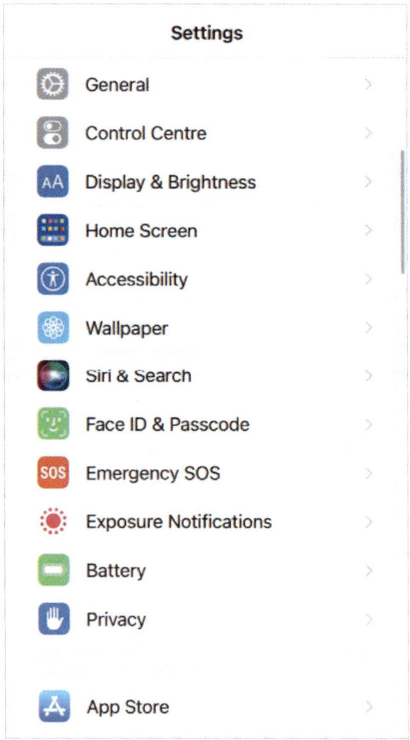

01 Go to Settings

Hit the Settings icon on your Home screen to go to the Settings section. The various options are grouped into categories that relate to the aspect of your iPhone that they apply to.

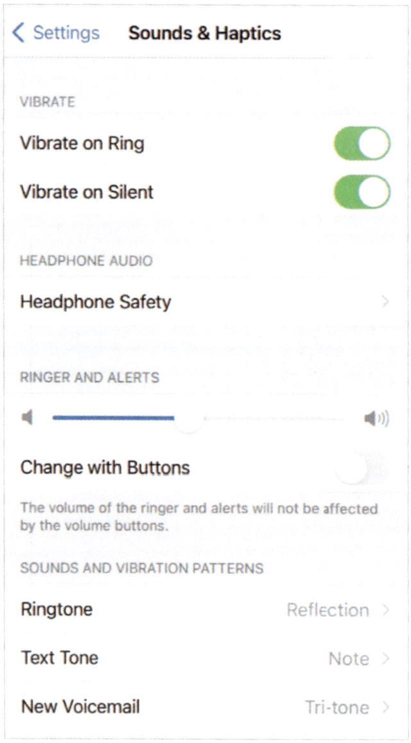

02 Customise sounds

Go into the Sounds & Haptics section in the Settings menu. Here you can set your ringtone tune and specify whether sounds should play when you get a new mail or voicemail.

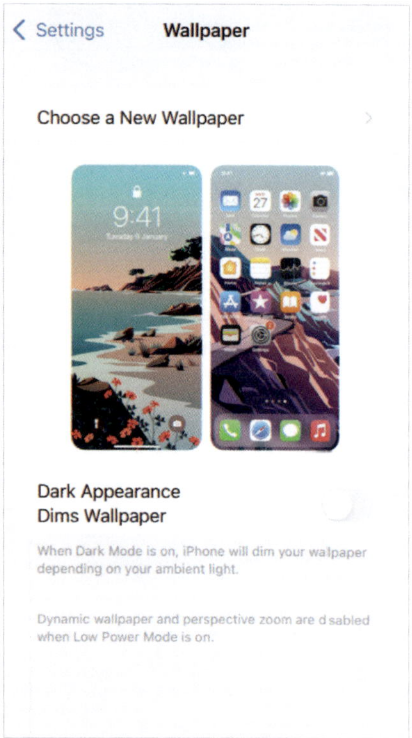

03 Change your wallpaper

In the Wallpaper area you can pick an image or use one from your photo gallery. You can also adjust the brightness of your phone and the text size by going to the Display & Brightness settings.

Manage your iPhone's settings

Most of your phone's functions can be customised to suit you

General settings
All the stuff that doesn't fit elsewhere, including serial numbers, data usage, Date & Time, software updates, CarPlay and more. This is a very important section to know about

Wallpaper
Alter the wallpaper and choose from Apple's images or any picture from your image library. Choose separate wallpapers for the Home and Lock screens if you like

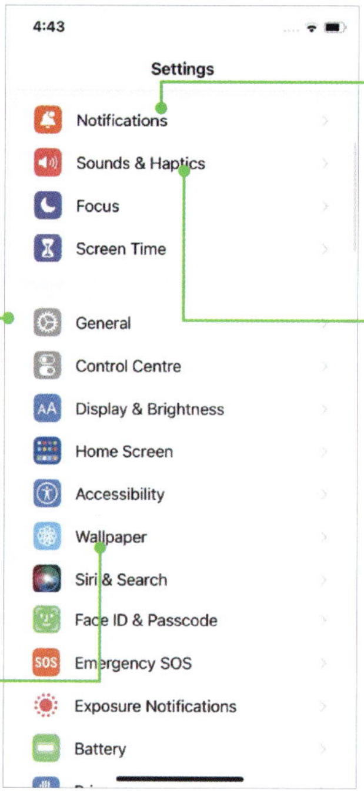

Notifications
iOS can push notifications to you when certain things are updated, such as new travel information. This notifies you without you having to manually check for updates

Sounds
Choose from the bundled ringtones or add your own unique ones, synced from iTunes to your phone. Also manage vibration and text alert tones, key clicks and other system sounds

Back up
When you connect your iPhone to your Mac or PC, it performs a backup of any changes, making applying your settings to a new iPhone very easy.

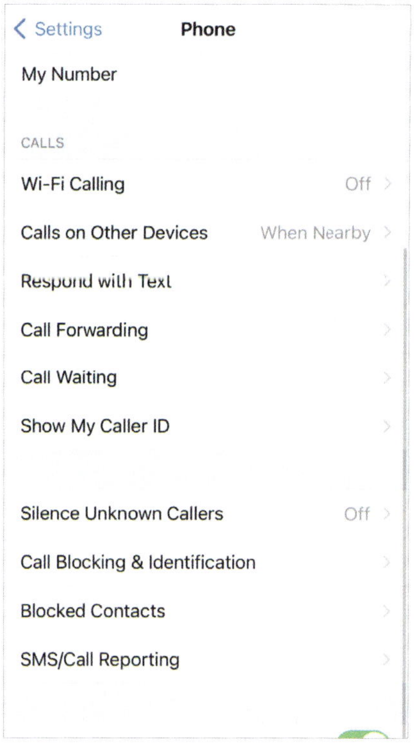

04 Phone section
This section displays your number and the ability to reply to a call with a pre-set or custom message. You can also activate a PIN code for your SIM card and assign SIM Applications.

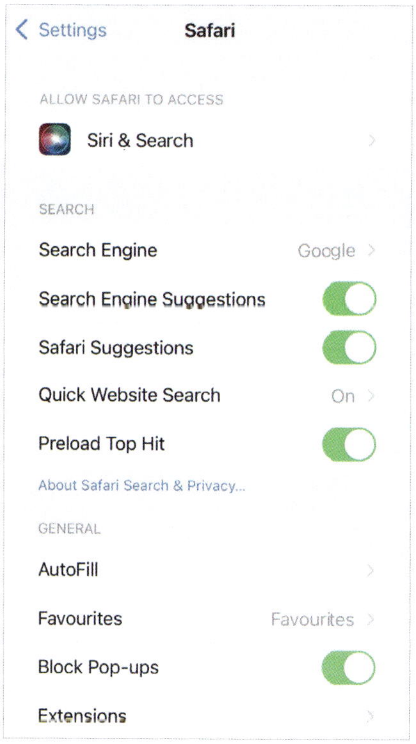

05 Set up Safari
By default, Safari blocks pop-up windows, which is handy. In this section you can also get it to autofill forms and passwords, which are stored securely in your iPhone's keychain.

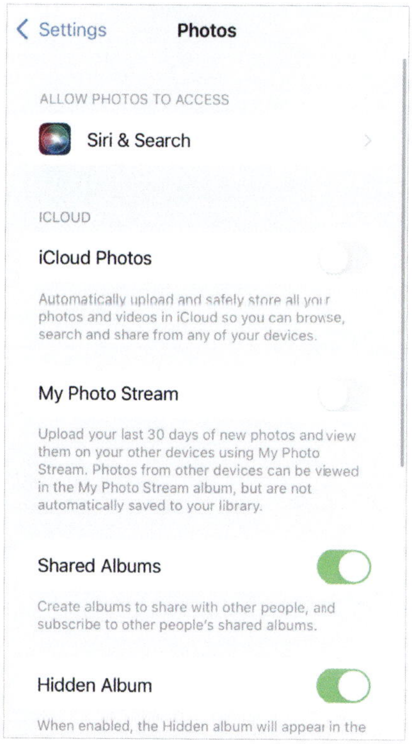

06 Photos section
In the Photos section you can enable iCloud Photo Library, Photo Stream and choose if photos are saved in full resolution on your iPhone. Selecting Optimise iPhone Storage can save memory.

Set up a Wi-Fi connection

There are a host of services on the iPhone, but you'll need to get online. We show you how to do just that

For all the iPhone's uses and versatility, it doesn't really come into its own until you have Wi-Fi access. Sure you can play games, write documents, send emails and even download the latest books, but you can't do the really big stuff if you don't have access to the Internet.

Internet access turns your iPhone from an impressive piece of expensive kit into something that opens up worlds. This is due in part to the magnificence that is the App Store, a virtual store that gives you access to literally thousands of different applications, ranging from popular games like Angry Birds and Real Racing to GPS devices, interactive encyclopaedias, word processing apps and much, much more.

Your iPhone is a portal to a host of fantastic services, but first you need to give it access to the Internet – only then do your options become plentiful. Follow our simple instructions in this tutorial and unleash the potential of Apple's powerful device. You won't regret it.

Settings Switch on Wi-Fi on your iPhone

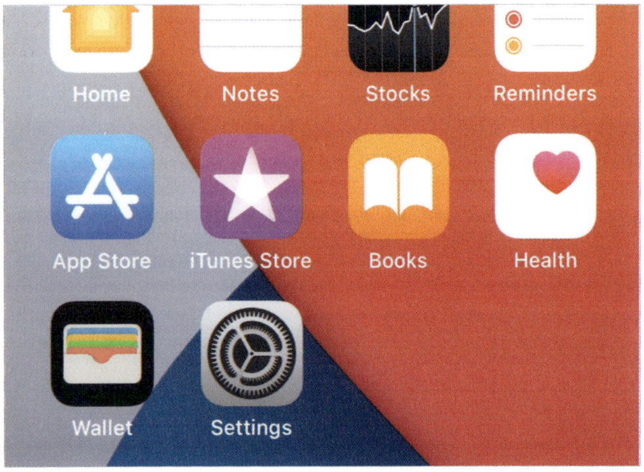

01 Getting started
Go to the very first page of your iPhone's Home screen and tap on the Settings icon to launch your Settings app.

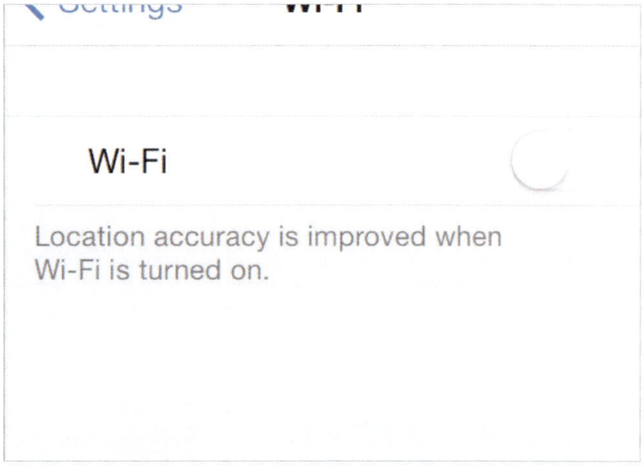

02 Locate Wi-Fi
After entering Settings, the second option down is Wi-Fi. Tap on it and then turn Wi-Fi on by sliding the button.

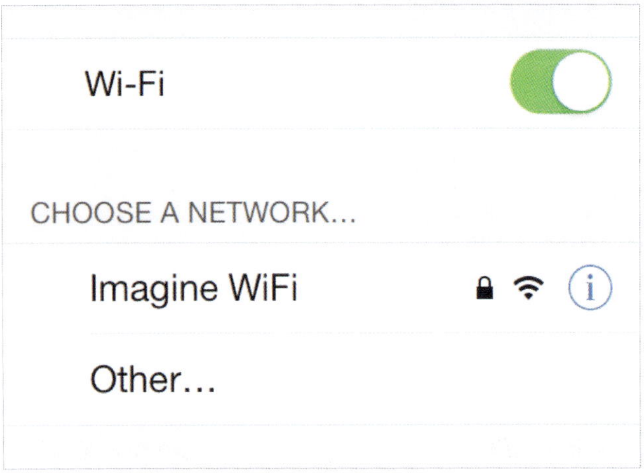

03 Find your connection
Your iPhone will now start looking for available connection points. Ignore Ask to Join Networks at the bottom and simply tap on your connection.

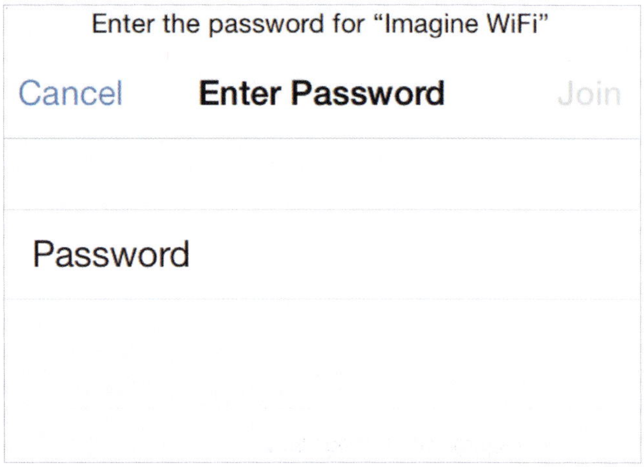

04 The final hurdle
You are now ready to enter your Wi-Fi password. Fill in your details on the keyboard. Once filled in, select Join and you're done.

Connect to 4G/5G

Access the Internet while you're away from home

4G and 5G are network services that allow you to access the Internet while you are away from a Wi-Fi signal, for example when travelling. It's an amazingly useful service, due to the fact that, as communication networks expand, online access is becoming more and more expensive. While public Wi-Fi spots are becoming more numerous – everywhere from pubs to McDonald's has them, and many are free – there can still be charges, in some hotels for example.

4G access comes with all iPhones and is enabled by default, which means you can access the Internet wherever you happen to be, whether there's a Wi-Fi signal or not. However, if you're going abroad or looking to save on battery life, it's a good idea to know how to tweak your settings, as you may want to avoid roaming charges or using up your data allowance unnecessarily. Follow these simple steps to set up your 4G or 5G and manage the way you access the Internet on your new device.

Settings Change your iPhone's 3G settings

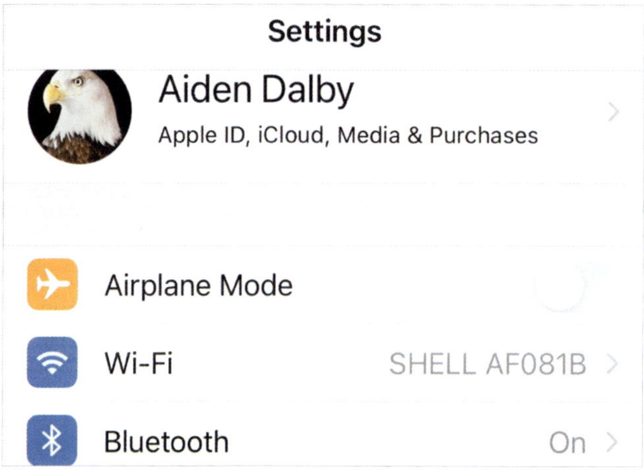

01 Find your settings
In the Settings app, look to the first section of options right at the top of the interface. You'll see one called Mobile Data, second from bottom; give it a tap.

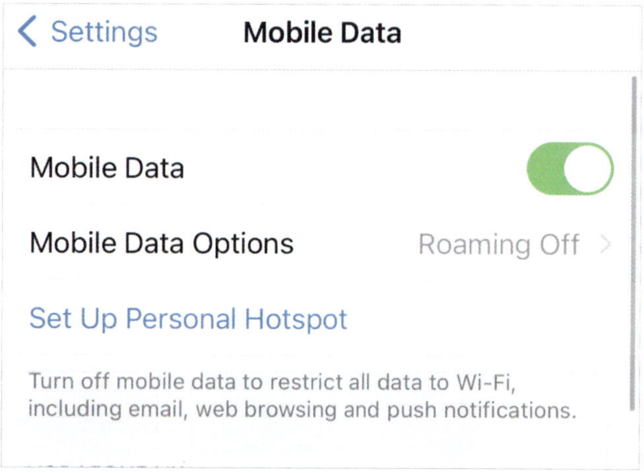

02 Network settings
You can toggle Mobile Data on or off at the top of the screen. Your battery will last longer with it off, but your data connection will be limited to slower speeds.

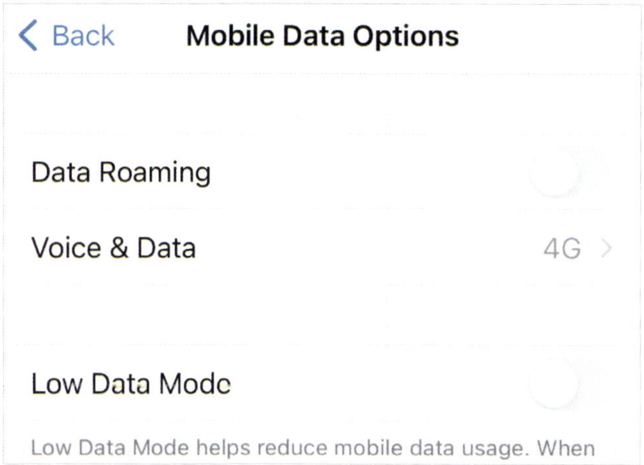

03 Data roaming
Using 4G while abroad can rack up massive charges, so the Data Roaming option is of interest to budget-conscious travellers.

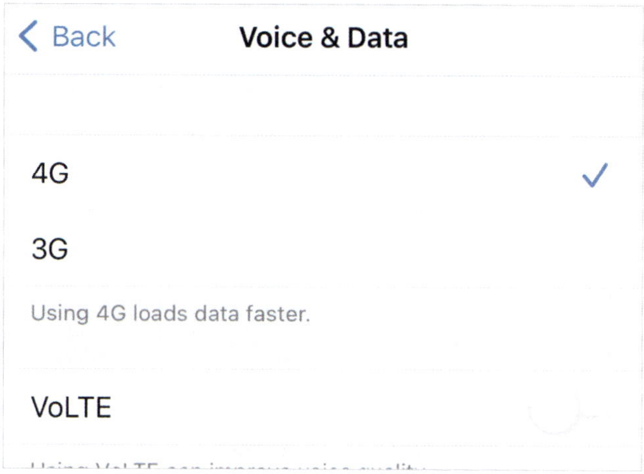

04 Mobile Data Options
You can switch between 4G and 3G by tapping on 'Mobile Data Options' then 'Voice & Data.' 3G uses less data than 4G but the battery life will last longer.

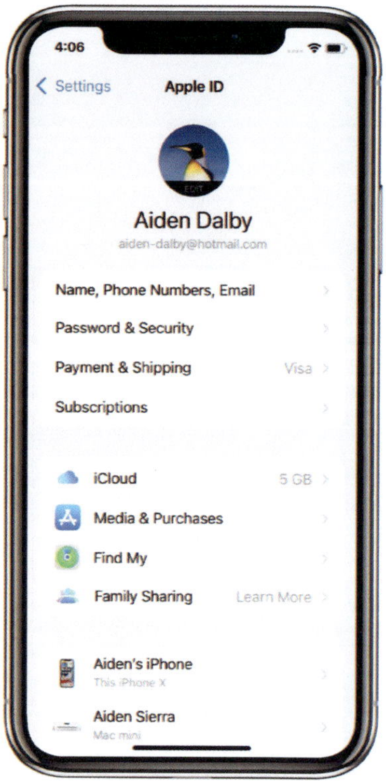

Set up your iCloud account

Use iCloud to move content wirelessly between all your Apple devices without having to lift a finger

Introduced to iOS a while back, iCloud is the perfect wireless solution for keeping all of your content up-to-date and synced between devices. Free to all users, iCloud automatically and securely stores your content (including music, apps, photos and documents) so that it's always available on your iPhone, iPad, Mac – whatever device you're using.

In the past iCloud was something separate from your Apple ID that you would need to sign into, and it had its own setup process. Thankfully all of that has since gone away, and you can easily turn on iCloud once you've signed into your Apple account. Also, more flexibility has been bestowed on the apps that you use to open documents created in other apps. You can also upload your entire Photo Library to iCloud Drive so that all of the photos you have ever taken on an iOS device are with you at all times. Read on to find out how to set up and use your iCloud Drive.

"iCloud Drive is a lot more versatile in allowing you to keep all of your files and documents safe"

Settings Set up your iCloud Drive

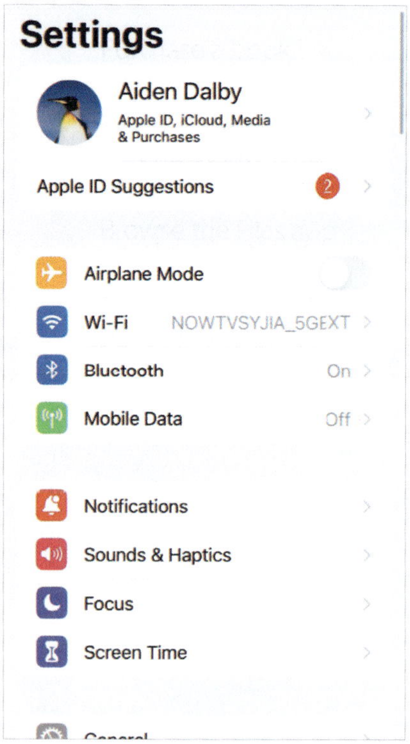

01 Launch Settings

There is very little tinkering to do to set up your iCloud. Once you've signed into your iPhone with your Apple ID you can access the iCloud options from the Settings app.

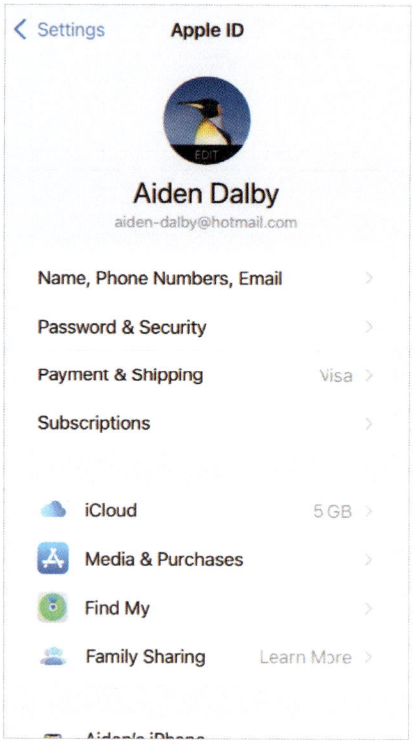

02 Sign in

At the very top of the Settings menu will be your Apple ID. Tap on this. If you haven't already signed into your Apple ID do so now and once inside, tap on the iCloud option.

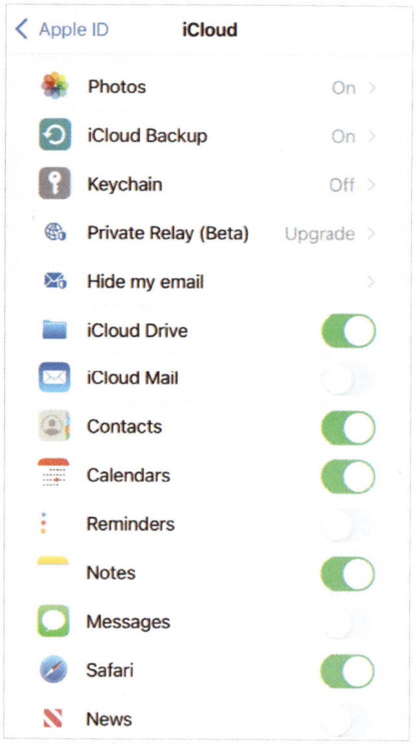

03 Activate iCloud Drive

You will see an option for iCloud Drive, tap on this and then you will be able to use iCloud for storing files and documents in apps such as Files. This is free and only takes a few minutes.

Your settings in iCloud

Activating and managing your free iCloud account is simple and easy

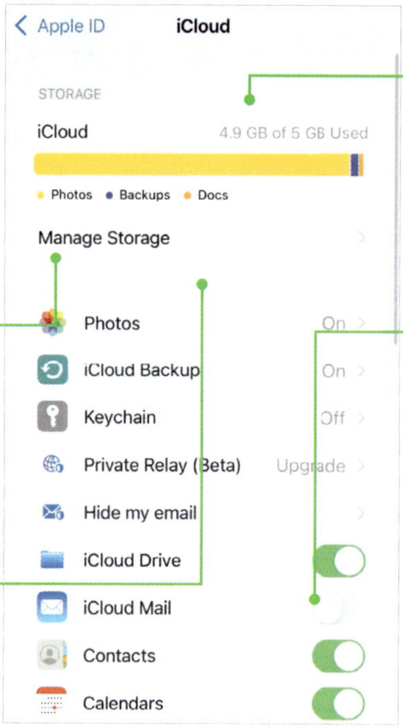

Your account

Your free iCloud is activated when you enter your Apple ID. This is the same email address and password that you use for other Apple services, like iTunes

Storage

You are provided with 5GB of storage space free with your iCloud account. If you find this to be insufficient, tap here to upgrade your capacity and see what other plans are available

Compatible apps

iCloud is integrated into a host of default Apple apps and works behind the scenes to make sure everything is synced across all devices

Family Sharing

Introduced with iOS 8, this feature lets family members share apps and help you monitor the apps that young family members use and even set up one central payment system for authorising and buying them

Back up your iPhone

In the iCloud settings you will see an option called Backup. Tap on this and ensure that the 'iCloud Backup' slider is moved to the 'On' position to automatically backup your data when the iPhone is charging, locked and connected to Wi-Fi. You can also tap on the Back Up Now option to instantly force your data to be backed up to your personal iCloud.

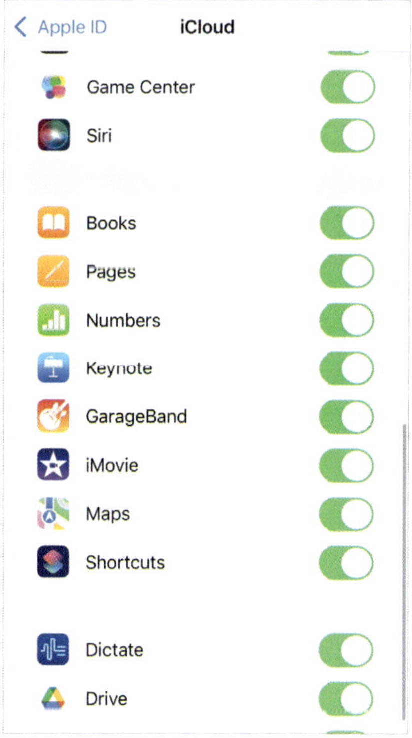

04 Activate apps

In the iCloud Drive settings you will see all of the apps installed on your device that use the service. Move the sliders to the 'On' position to enable files for these apps to be stored in the cloud.

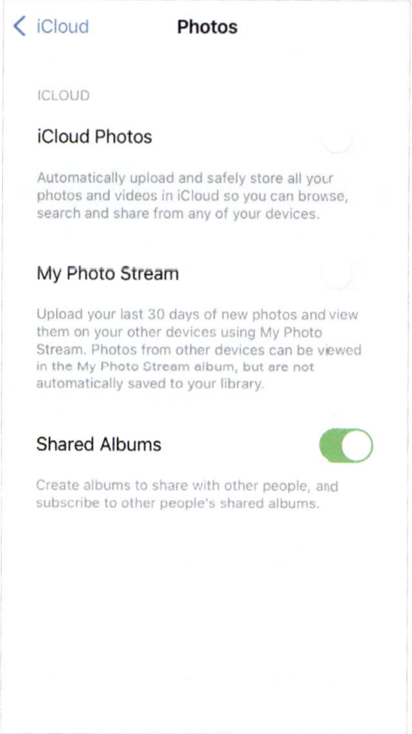

05 Tailor your options

Go back to the iCloud settings and select which apps are synced with your iCloud and kept up-to-date across all of your devices. Go to Photos to turn on your iCloud Photo Library.

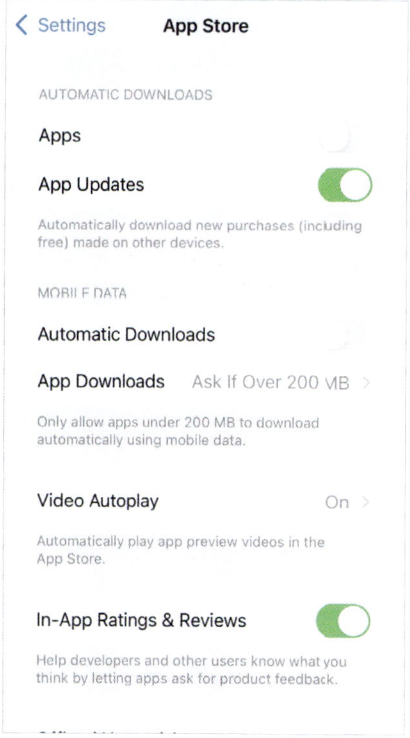

06 Automatic downloads

To get the most out of iCloud, click on the App Store or Music section in Settings and then turn on Automatic Downloads to download new purchases from other devices.

Sync your apps with iCloud

Your iCloud can work wonders behind the scenes to wirelessly move your documents between devices

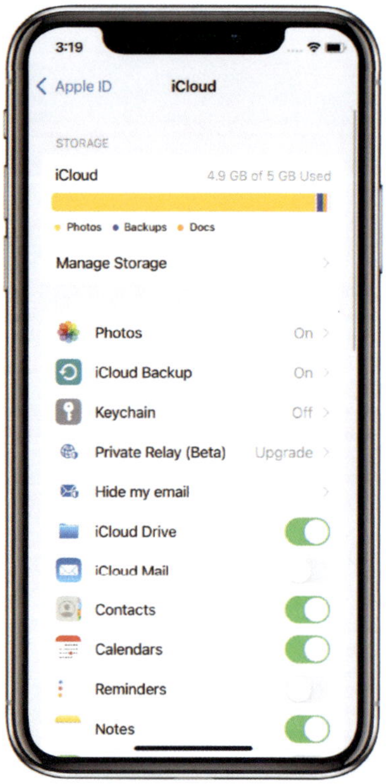

Now that you've signed into your iCloud account using your Apple ID, you'll have much more freedom to sync your files and access them in other apps. Essentially, iCloud still works in the same way across every app (you can enable the iCloud function with a range of apps in Settings), in syncing files and making them accessible across all of your iOS and macOS devices. iCloud Drive is a bulkier beast to which you can save all of your files and documents and make them available anywhere. You can turn on iCloud Drive from the iCloud options in the Settings app.

With apps such as Pages, Numbers and Keynote as well as a wide range of third-party apps you can now enable iCloud when starting up the updated app, which means all of your files and documents will be uploaded to your iCloud Drive and made accessible, not only on your other devices, but in other apps too via the dedicated Files app. Read on to learn more.

"iCloud Drive is a bulkier beast to which you can save your entire Photo Library"

Settings Which apps should you activate?

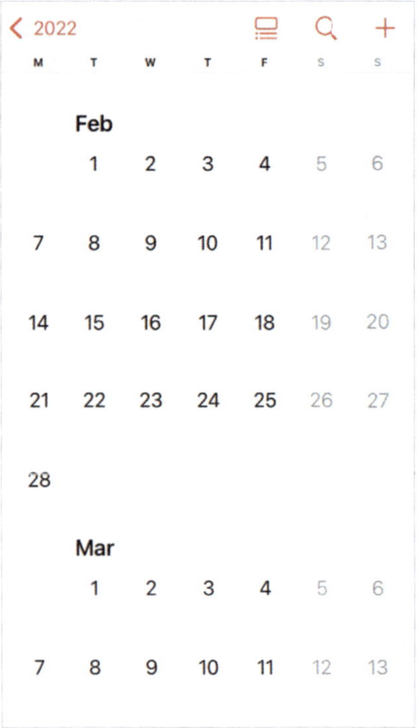

01 Mail, Contacts & Calendars

If you lead a busy lifestyle and use multiple devices, then you should enable all three of these options so that calendar events, contacts and emails are synced and updated.

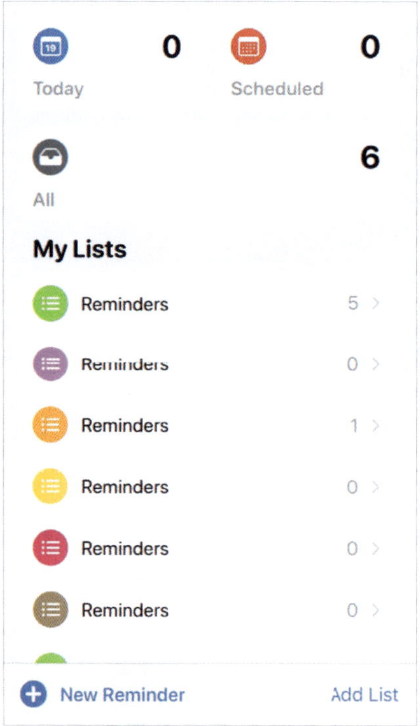

02 Reminders

Reminders is a simple app that lets you make 'to do' lists and set reminders. By activating it in iCloud all of your devices will remind you, not just the one you set the reminder on.

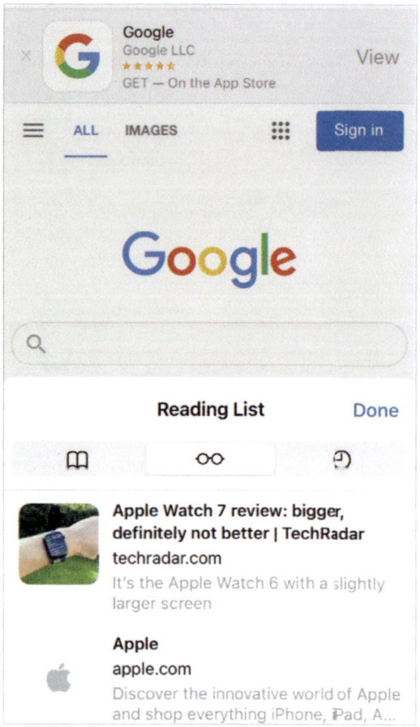

03 Safari Bookmarks

Enabling Safari ensures that all of your bookmarks, Reading List items and open tabs are synced across all of your iOS or macOS devices, ready to access wherever you are.

Activate iCloud with apps

Syncing apps to iCloud is as easy as moving sliders with your fingers

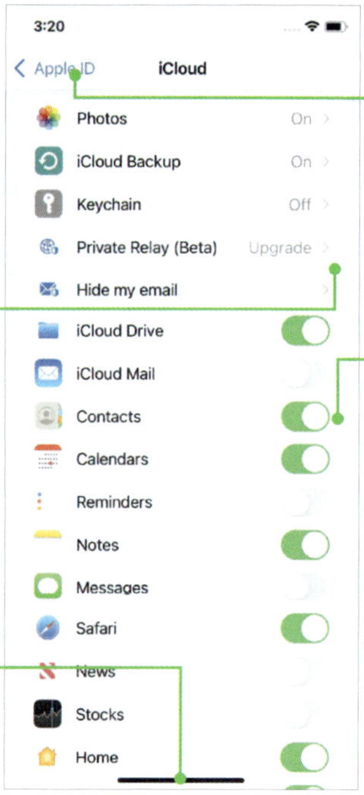

Your account
You can view your iCloud account information by tapping on the Apple ID section at the top of the main Settings screen

Extra options
Some iCloud functions have extra screens that you can access by tapping the arrows. There are no additional options associated with them, just more detailed explanations

Enabling apps
To utilise the iCloud features of your apps to sync info and data across devices, simply move the slider of each app to the 'on' position

Delete account
If you are upgrading your device and wish to disassociate your current iPhone with an iCloud account then simply tap the Sign Out option at the bottom of the list

Peace of mind
If you misplace your iPhone, worry not. As long as you enable the Find My iPhone option on the iCloud Settings screen, the position of your device will be stored in your iCloud and can be accessed from any computer web browser by going to **www.icloud.com.**

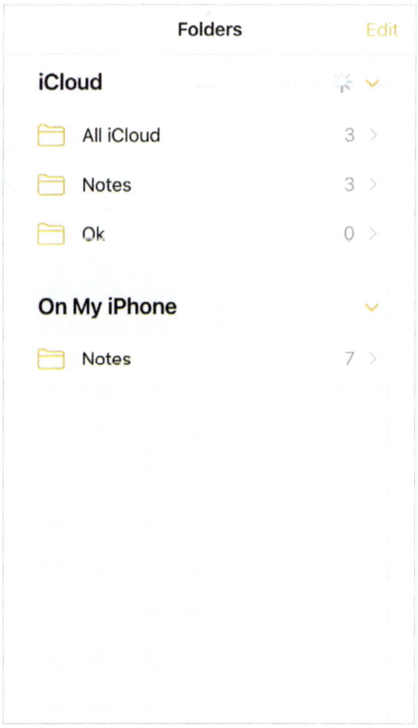

04 Notes

Moving the Notes slider to the 'on' position will ensure that any notes that you make through the Notes app on one device will be synced and made available to read and edit on other devices.

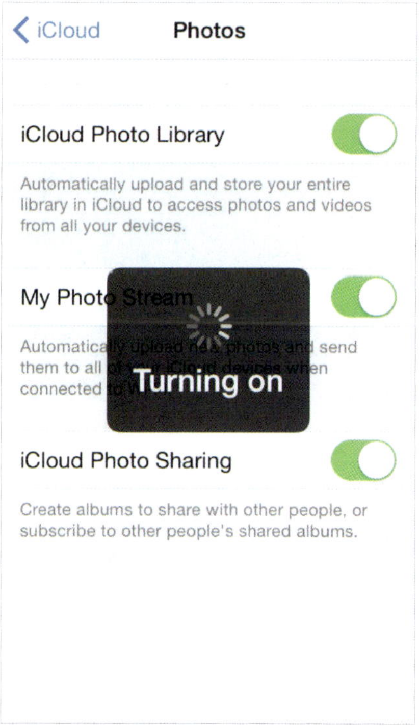

05 Photo Stream & Library

With Photo Stream enabled, any photo you take on your iPhone will be sent to all of your other iOS and Mac OS devices. Enabling Photo Library will store your entire collection in the cloud.

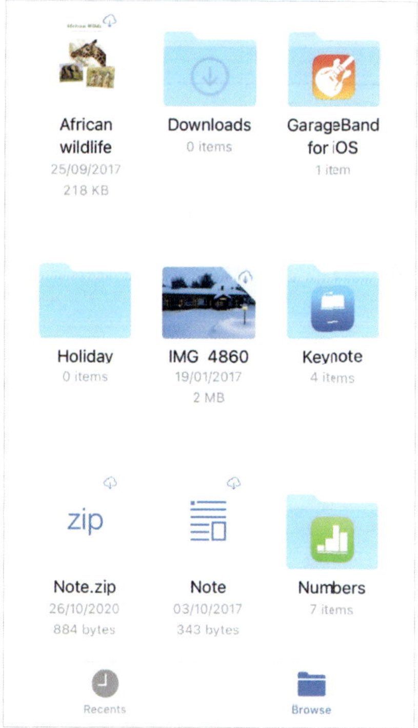

06 Documents

With iCloud Drive, all of your documents and files from various apps are stored in the cloud and made available to other apps via the Files app. Just enable iCloud when starting up a related app.

Use iCloud to back up your iPhone

With iCloud, you can back up all of your important iPhone data to your own virtual hard drive

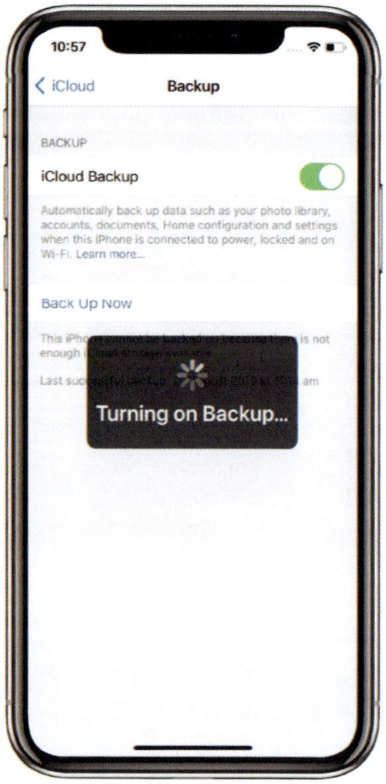

Your iPhone is like a bank vault where all kinds of data is stored. So what happens if your iPhone gets lost or breaks? Nothing. Thanks to iCloud, all of your data is backed up automatically and kept safely in your cloud storage space. When your iPhone is connected to a power source and a Wi-Fi network, all of your media, settings, app data and messages are backed up.

When you set up a new iOS device or need to restore the information on the one you already have, iCloud Backup does all the heavy lifting. All you have to do is ensure that your device is connected to Wi-Fi, enter your Apple ID and all of your important data will appear on your device.

The benefits of using iCloud are vast, and the way in which it goes about its business in the background without you having to worry about prompting it is just another prime example of how Apple is striving to make your life easier, and keep you connected with each device you own.

"iCloud is another prime example of how Apple is striving to make your life easier"

Settings Activate iCloud and back up your data

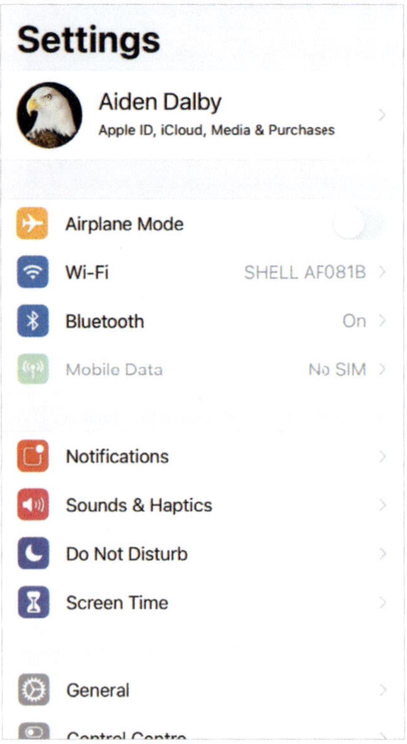

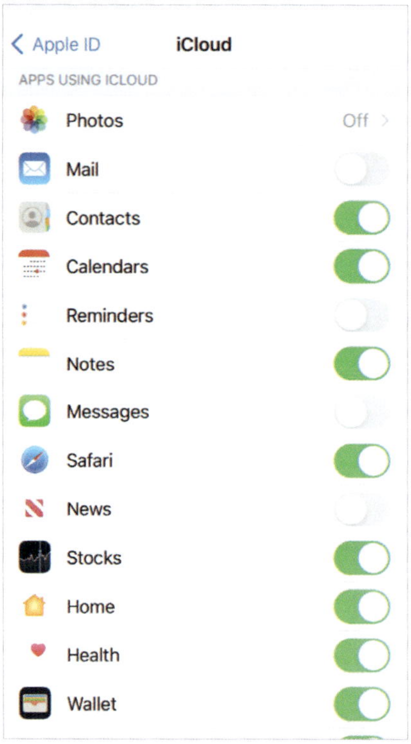

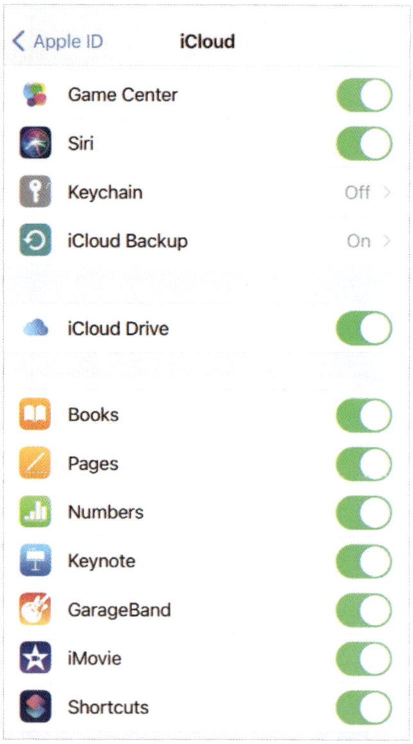

01 Go to Settings

On your iPhone, launch your Settings app and then tap on your account. You will see the iCloud section, so tap on that to reveal the extensive list of options associated with it.

02 Sign in

You may need to sign into your iCloud using your Apple ID (email address and password), but once done you will be able to start tailoring the various iCloud options to suit your personal needs.

03 Set up

Once your personal iCloud has been set up and you have selected which apps you would like to sync, tap on the iCloud Backup option near the bottom of the list.

Back up with iCloud

Once activated, your iCloud will back up your stuff automatically

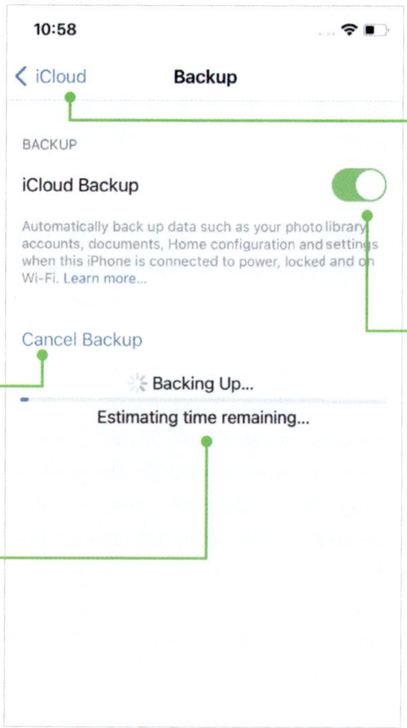

Manage your space
You are able to see how your free 5GB allocation of cloud storage space is used by tapping Manage Storage in the main iCloud Settings screen. Tap here to go back

Backup
When the iCloud Backup option is turned on, all settings, documents, media and photos will be automatically backed up when your device is connected to a power supply and Wi-Fi

Instant backup
You are able to back up your data by tapping the Back Up Now option. Do this if you change your settings or buy new media

Progress bar
When your device is backing up, a progress bar will appear that shows how far along the process is. You can cancel the backup at any time

Manage your storage
By tapping on the Storage option in the iCloud Settings screen and then going to Manage Storage, you'll be able to see exactly how your iCloud is used with individual breakdowns of how much of your free 5GB of space your apps use. If you find that you need to purchase more space then you can do so through this screen by tapping 'Change Storage Plan'.

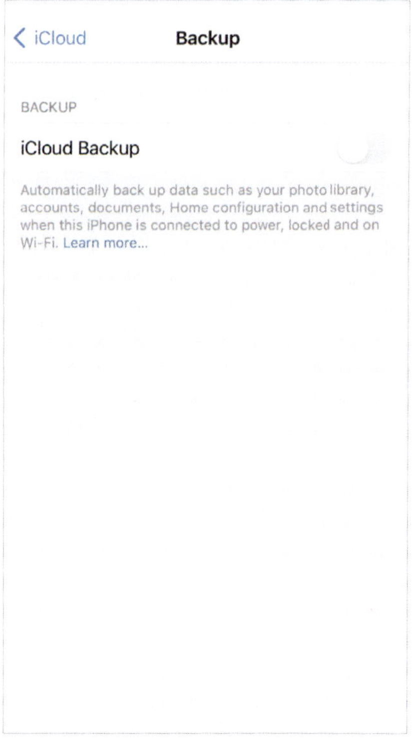

04 Turn on backup

On this screen you will be able to monitor and manage your iCloud storage space, but more importantly, you will see an option called iCloud Backup. Ensure that the slider is moved to On.

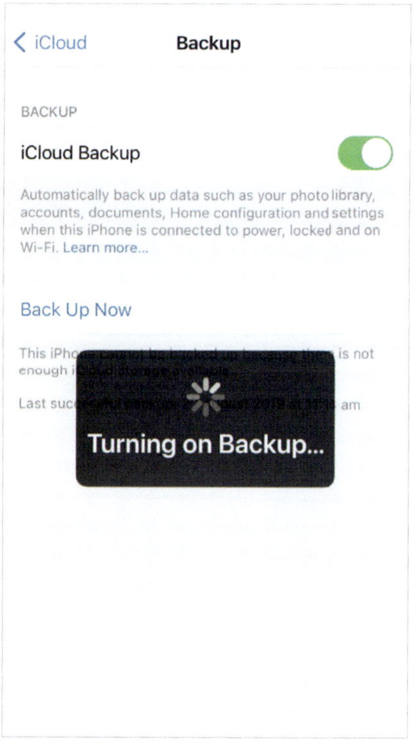

05 Wait for the activation

You'll be presented with a message saying that your iPhone will no longer sync to iTunes when connected to your computer, so tap OK and wait for a minute while the Backup feature is activated.

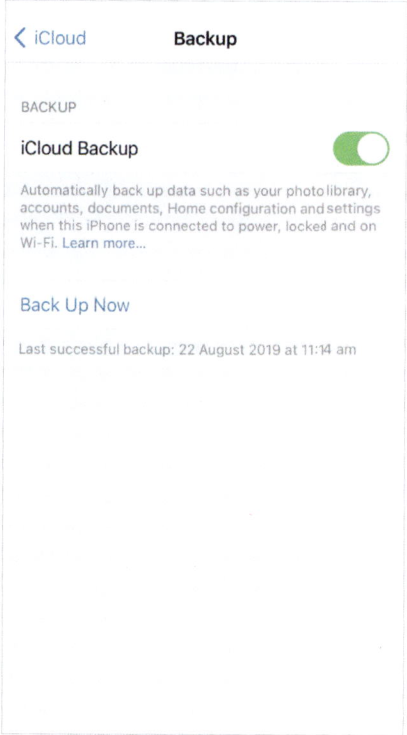

06 Start backing up

Your iPhone will now back up when connected to a power source and a Wi-Fi network, but you can perform the backup whenever you want by accessing this screen and tapping Back Up Now.

Set up a Passcode Lock

Protect your phone from prying eyes by setting a passcode that'll ensure that only you have access to everything on your device

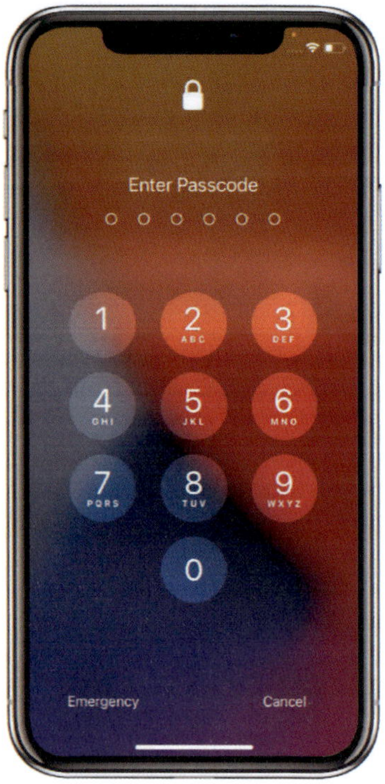

These days, people can store a lot of sensitive data on their smartphones – whether it's log-in details for online accounts, private text messages or documents containing private information. If someone gained access to your phone, there may be a lot on there that you don't want them to see. On a more basic level, you may also have friends who think it's funny to send messages from your phone when you leave it unattended for a few minutes. By setting a passcode, you can ensure there is at least some protection in place for the contents of your phone.

Working just like a PIN, you can set a passcode that only you know, and every time you turn your phone on or unlock it, you'll have to enter this code. Choosing a memorable code is imperative, but there are a range of settings to allow you to personalise your security to the desired level. The option for more complex passcodes was introduced in iOS 9, so you can tailor the Passcode options to suit you. You can even set your phone to erase if the passcode is incorrectly entered enough times in a row. No longer will you need to worry about your phone's data falling into the wrong hands.

"No longer will you need to worry about your phone's data falling into the wrong hands"

Settings Turn on Passcode Lock

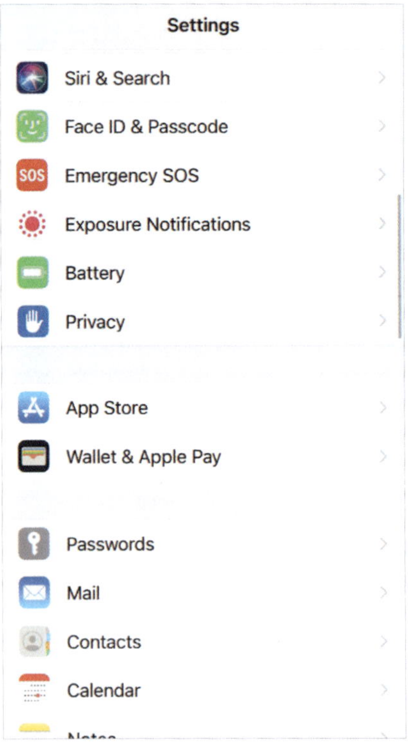

01 Through Settings

You will be asked to set a passcode when setting up a new iPhone or updating to the latest version of iOS, but if you want to do it later, launch the Settings app and scroll down to Face ID & Passcode.

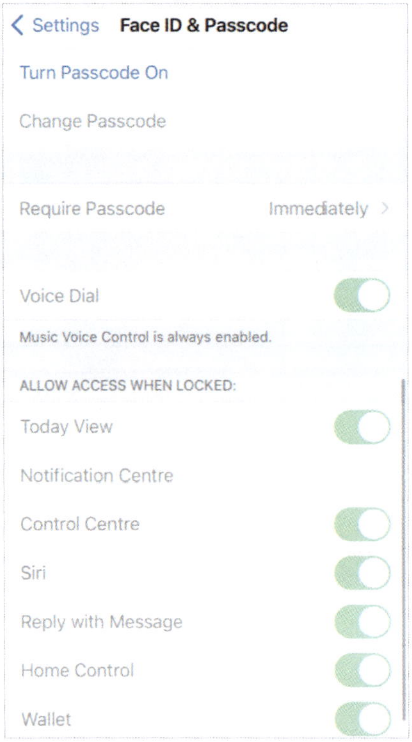

02 Passcode Lock

Within the Passcode settings, you'll see the display of options shown above. Currently, there is no passcode set on this iPhone. In order to change this, tap on the 'Turn Passcode On' option.

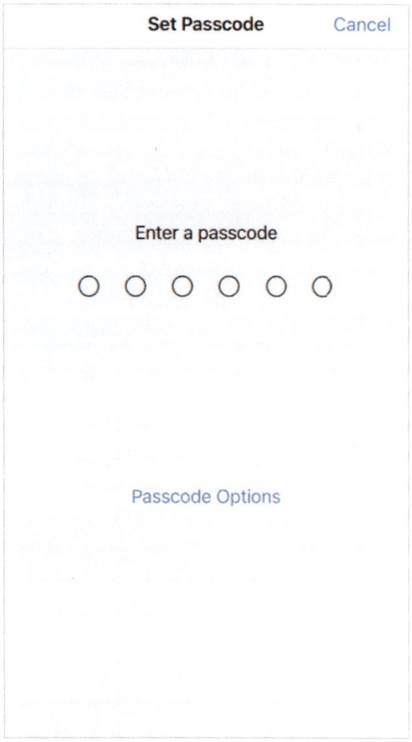

03 Enter a code

You will be asked to enter a six-digit passcode. Make sure to choose something you'll remember, but nothing too obvious that people will guess. Re-enter the code to confirm when asked.

Secure your iPhone

View your passcode options

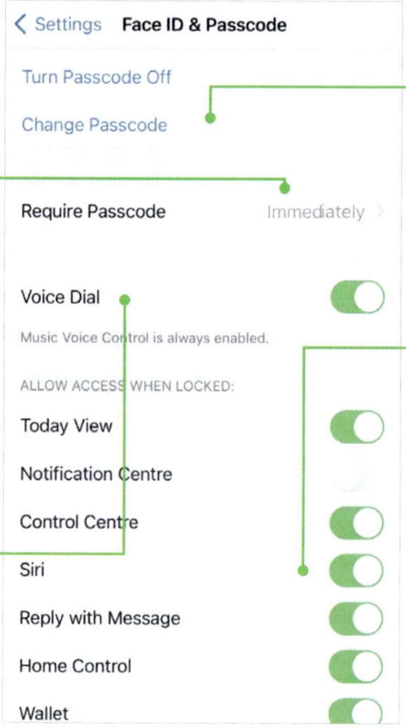

More options
Tap this option and then choose 'Passcode Options' to change the passcode to something more complex by choosing the Custom Alphanumeric option

Waiting time
Tap this option to select how long you'll have to wait to enter your passcode after locking your phone – from immediately up to four hours

Turn it off
These let you control whether or not Siri, Passbook and the Phone app's 'Reply With Message' work if your iPhone is locked

Voice Dial
Flick this switch to select whether or not you want the Voice Dial function to work without entering a passcode

Erasing data
Within the Face ID & Passcode settings page, you'll find an option to Erase Data. If you're concerned about losing your phone or have very sensitive data that you don't want falling into the wrong hands, drag this slider to 'On'. This means that if a passcode is incorrectly guessed ten times in a row, your phone will be wiped.

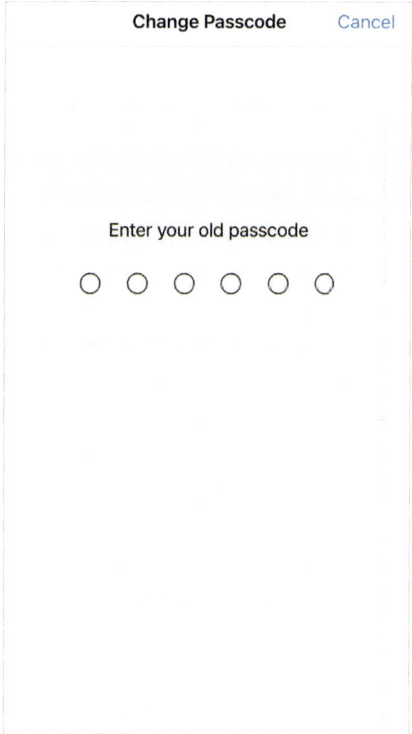

04 Extra security

To go about making a more secure passcode, tap on the 'Change Passcode', re-enter your six-digit passcode and then tap on the 'Passcode Options' when asked to enter your new passcode.

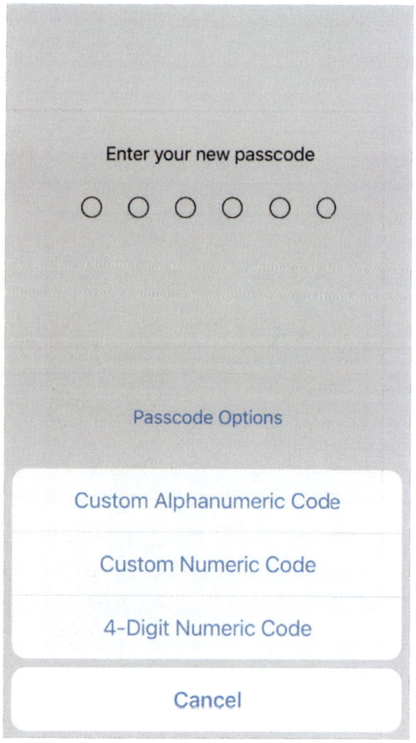

05 Select new passcode type

You will now be presented with three options: Custom Alphanumeric Code, Custom Numeric Code and 4-Digit Numeric Code. Choose the first option for maximum security.

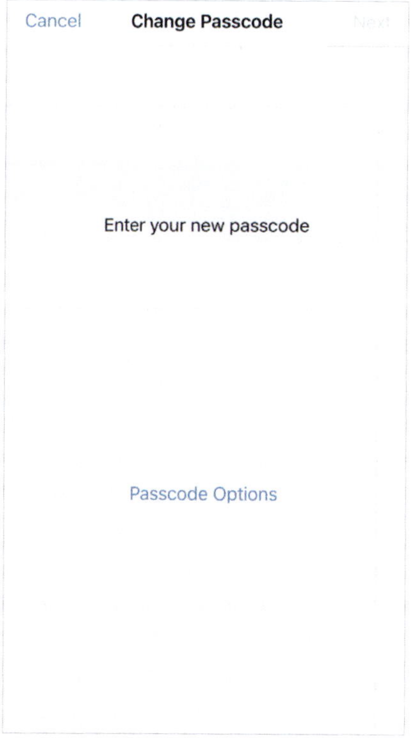

06 Enter new code

You can now use any of your iPhone keyboards, including the full QWERTY layout and symbols, to create a passcode for your iPhone that's much harder to crack.

Customise iPhone Face ID settings

Set up your iPhone to recognise your face for added security – it's much more secure than a passcode

Face ID ticks all of the boxes for a facial recognition system and works exactly as you would want it to. It's quick, easy to use and secure, but you still have many options that you can tweak to make it work in the way you want it to. For the majority of these settings, you should only need to set them up once and you will then enjoy greater unlocking speeds, more secure app purchases and the reassurance that the technology is working exactly as it should. We will show you how to enable the feature to your iPhone and also how to choose when Face ID should be used. Given that it's more secure than any password, we suggest enabling it in every available area. It takes a few minutes to complete the setup, but the time savings in the future are immeasurable compared to what you will need to do initially. If you have a secure Face ID device, it means that nobody should be able to access your data under any circumstances, but the requirement to still keep your passcode safe and secure is important. Face ID is reassuring, secure and extremely easy to use.

> "It takes a few minutes to complete the setup, but the time savings in the future are immeasurable compared to what you will need to do initially"

Settings Enable Face ID

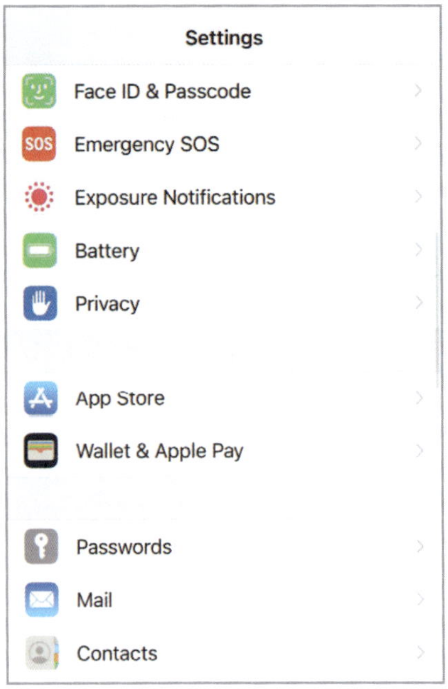

01 Locate the settings

Open the Settings app on your iPhone and look for the 'Face ID & Passcode' option. You will likely have to input your passcode before you are able to access this screen.

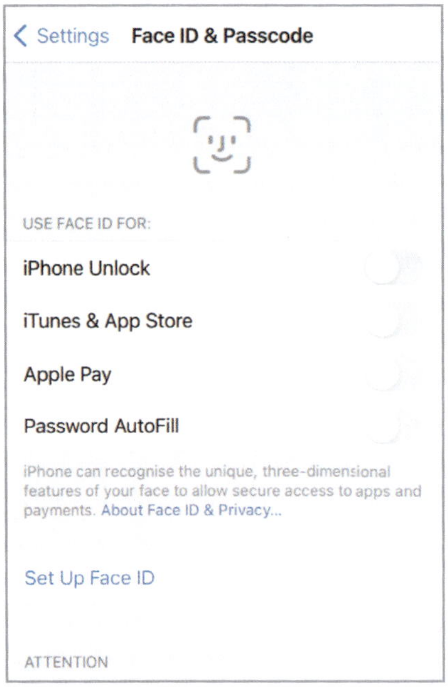

02 Set your options

At the top of the screen you will see options to set up Face ID for unlocking your iPhone, Apple Pay, Password AutoFill, and making both App and iTunes purchases. Tap what you need.

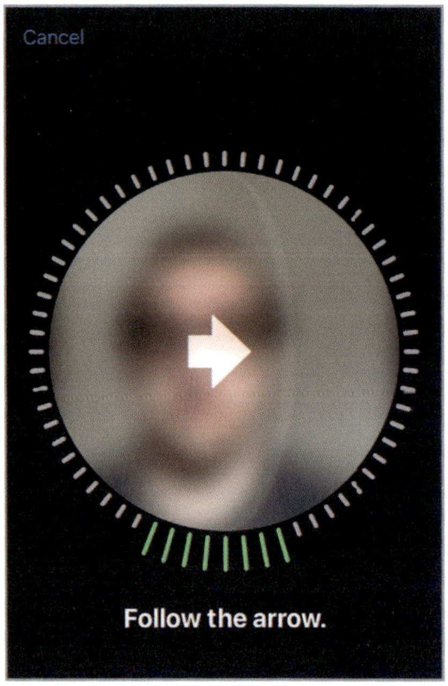

03 Follow guides

Before you start, a screen will appear telling you where to position yourself and what you should do so the camera can record your Face ID. These guides will also appear when capturing your face.

Multiple Face ID settings

Set up Face ID to work for you

Seriously clever technology
The technology behind Face ID is very clever and your personal data is only ever stored on a secure section of the iPhone hardware

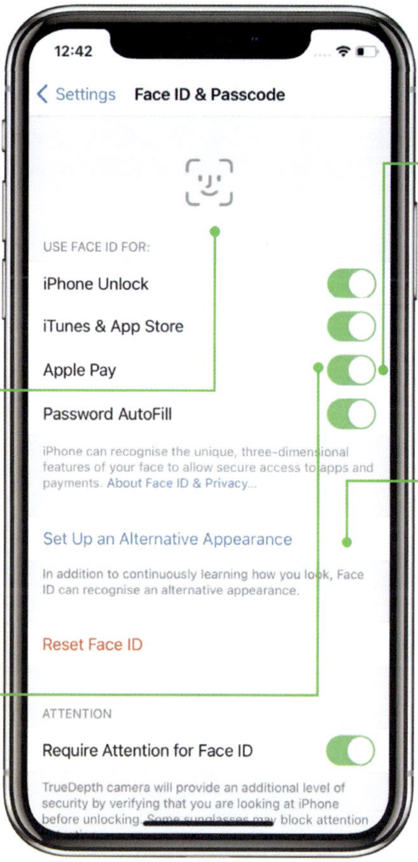

Secure payments
You can also use Face ID to enable secure Apple Pay payments and to avoid having to input a password when purchasing apps

Alternative appearance
You can use this option to set up a second person's Face ID so they can unlcok the iPhone as well

Quick unlocking
Your iPhone will be unlocked almost immediately thanks to the speed of Face ID. Just press the side button and the camera will begin to scan your face

How does it work?
The Facetime camera projects 30,000 invisible dots onto the user's face and creates a facial map to determine their identity. Apple claims that the odds of someone else being able to unlock your device are 1,000,000 to 1. The software can also adapt to recognise your face even if you make changes such as growing a beard.

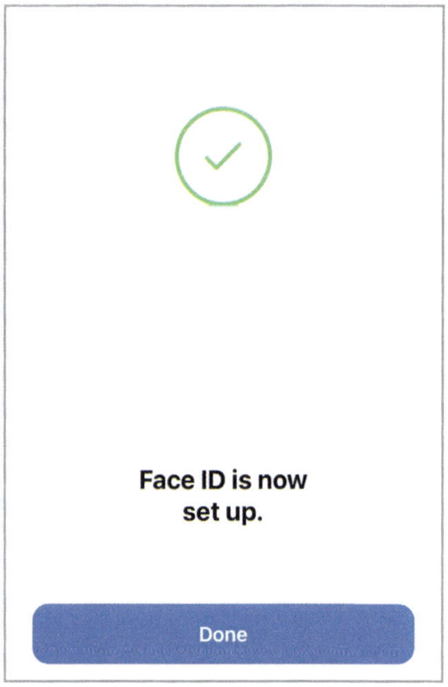

04 Setup complete

Follow the instructions carefully and you will eventually reach the completion screen. It takes some time, but it also saves a lot of time in the long run.

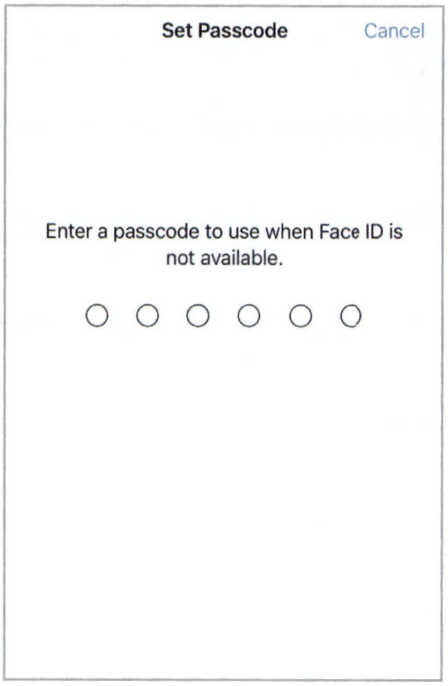

05 Use a passcode

You will also need to set a passcode that you have as a backup just in case the Face ID sensor fails. It will still be needed when powering on an iPhone, so make sure it's memorable.

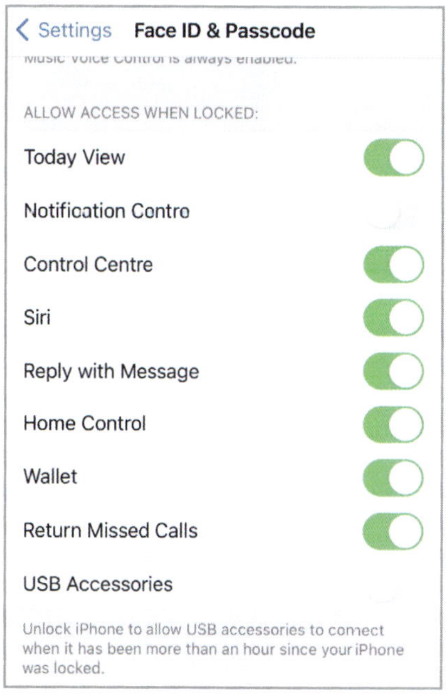

06 Access when locked

Further down this settings screen you can also allow access to other secure features, even when the iPhone is locked. These act like quick shortcuts from the lock screen.

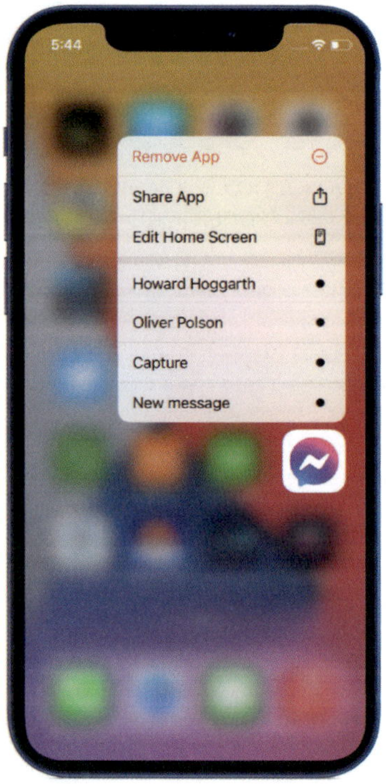

Understand 3D Touch

You may be thinking 3D Touch is a feature you can live without in your iPhone experience, but once you start using it you'll soon realise just how useful it is

At first 3D Touch may just sound like a gimmicky feature that may prove useful from time to time, but it is only when you start using it every day that you realise how much it can transform your user experience. Its incorporation was the first major redesign of the iPhone interface since the original phone was released, and it could be described as a right-click mouse button – but it actually goes much deeper than you may expect. It works by sensing the strength with which you press the screen, and can then feed back results depending on the degree of pressure detected.

A firm but light touch lets you peek at emails and other content, and an additional, harder touch will fully open the content you are viewing. In some apps, new options will appear when you press hard, and from the home screen you are able to press hard to see immediate controls such as Take Selfie in the Camera app.

We will guide you through the main ways in which you can use 3D Touch and explain where the feature is most useful, and it is quite possible that we will see many more uses appear as third-party developers take advantage of the system. 3D Touch is an efficient and clever time saver.

"It works by sensing the strength with which you press the screen, and can then feed back results depending on the degree of pressure detected"

3D Touch Learn how to work 3D Touch

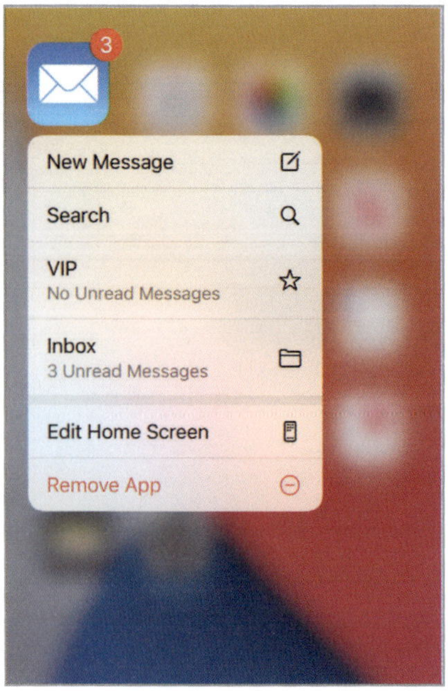

01 Press an icon

Press the Mail icon on the home screen and you'll see some options appear. Tap the Inbox option at the top to be taken straight to your latest emails.

02 Preview an email

Press any email in your inbox and you will see it displayed in a small window. You can now preview it and decide whether to fully read it or not.

03 A harder press

If you press on the email again and it will fully open for you to read it as normal. It's a very useful time saver for those with busy inboxes. There are other options such as Reply and Forward.

04 Peek at photos

When scrolling through a long list of your photos, you can use the same quick peek movement to quickly look at a photo and then move on to the next one.

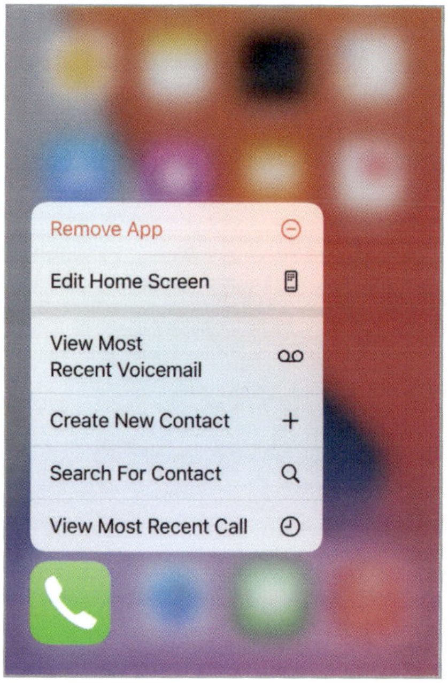

05 Quick calls

By pressing on the Phone icon you can find a shortcut to your most recent calls, search your list of contacts, create a new contact or listen to your most recent voicemails.

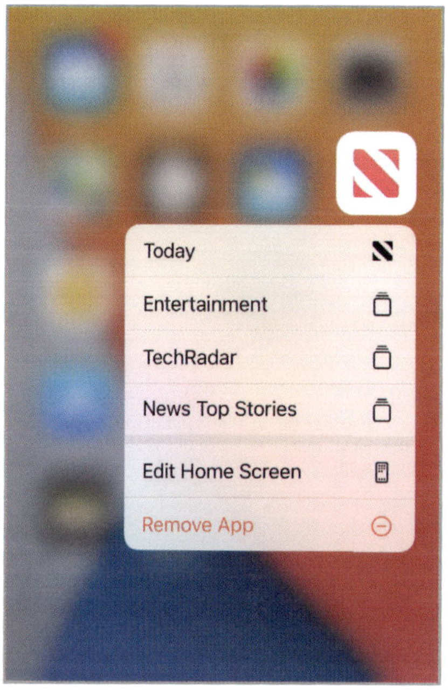

06 Access options faster

With some apps, such as News, you can press hard on their icon to find shortcuts to specific parts of the app. With News you can access your favourite news sources more quickly using 3D Touch.

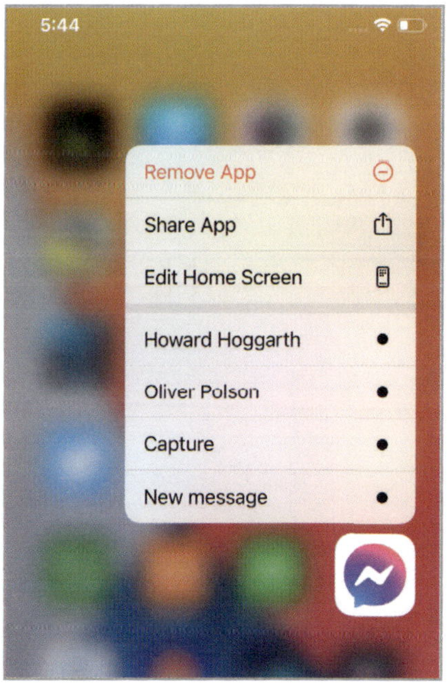

07 Third-party touches

App developers have the ability to include special commands that are specific to their apps. These can save you a lot of time with every use.

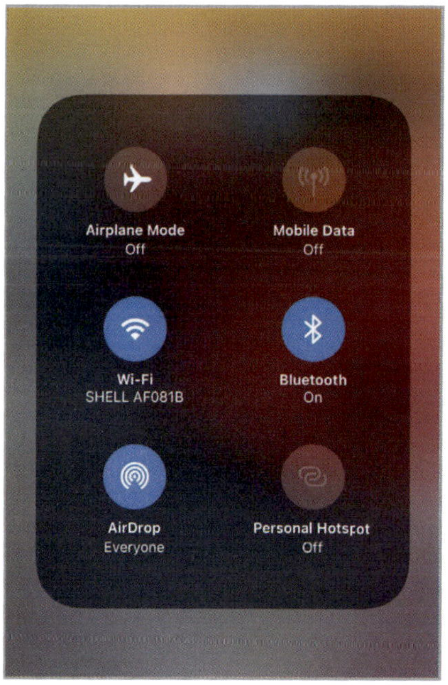

08 Control Centre

When viewing the Control Centre you can use the 3D Touch function to access more detailed controls. Almost every button in the Control Centre has a 3D Touch function.

09 Lock screen fun

You can select a live photo to be used as your phone's lock screen image, which you can press to animate. It's a fun feature that adds a touch of personality to your iPhone.

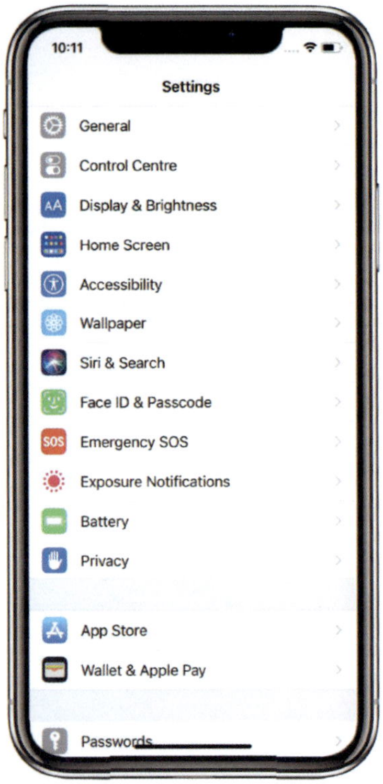

Take control of your privacy settings

Keep control of your privacy by monitoring the type of actions your phone's apps can and cannot perform

In today's digital age in which personal details are moved around at a moment's notice and critical information is bounced from one server to another at the flick of a switch, it is very important to keep control over your personal data. Think about it. Your iPhone contains a gamut of personal information. Just how many emails and texts do you think you send and receive during a typical week? What addresses do you store on your phone? Do you use your phone for personal banking?

First things first, you should always make sure to set up the Find My iPhone app, which is supplied by Apple for free and knows what to do if your iPhone is lost or stolen. But, thanks to features that were introduced by Apple years ago, there are now other ways to control the privacy of your data.

In one notable example, you may or may not be aware of the controversy surrounding apps like LinkedIn and Path, which were caught uploading information from users' address books to their servers without informing users. It caused a mini scandal. Features in iOS can prevent this from happening by allowing users to control which apps have access to what kind of data on their iPhone. Here's how.

"Features in iOS can prevent this from happening by allowing users to control which apps have access to what kind of data on their iPhone"

Settings Secure your iPhone's privacy

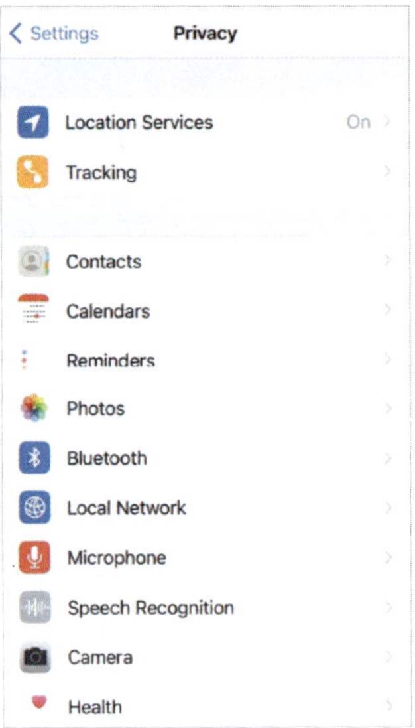

01 Open Settings

The Privacy settings can be found in the Settings app. They include Location Services, Contacts, Calendars, Reminders, Photos, Bluetooth Sharing, Twitter and Facebook.

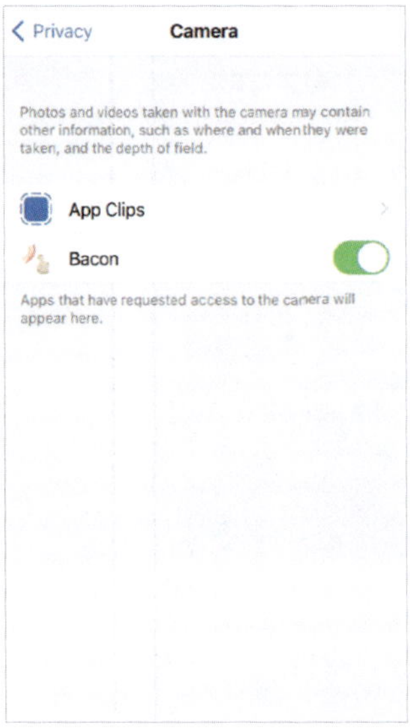

02 Seeking permission

Navigating to each of the sections gives you a suitable list of apps that have requested access to various other apps on your device. You can enable or disable access via the toggle.

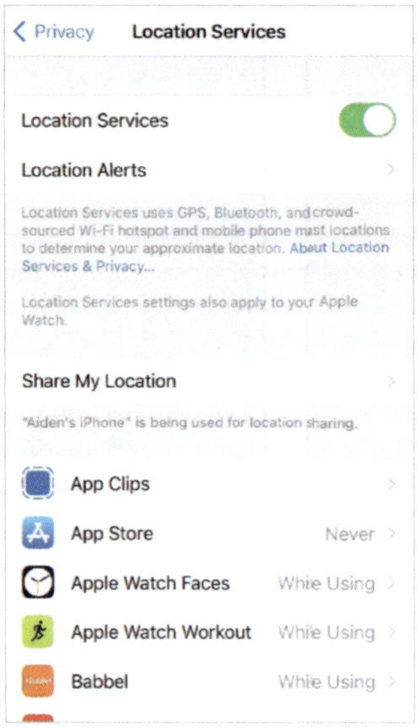

03 Location Services

When you enter this area, you will be told which apps have requested your location. More than that, however, you will also be able to see which app has used Location Services recently.

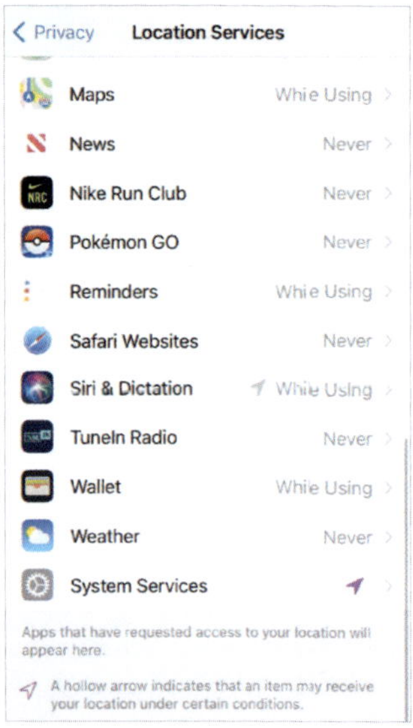

04 Arrow indicators

A purple arrow means that app has recently used your location. If it's grey then it was used within 24 hours, and an outlined icon means it will use your location under certain circumstances.

Maintain safe phone use

Keep an eye on your iPhone's activities to prevent your data from falling into the wrong hands

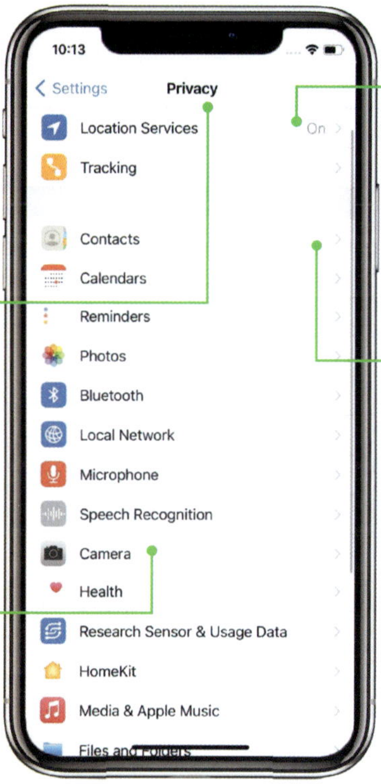

Privacy section
Most of the issues relating to privacy on the iPhone are stored within this section. There are other areas of note to be found within the Advertising section

Camera
When you take a picture or shoot a video, the app may also record when and where it was taken from. So be aware of what apps have access to your photos

Your location
The Location Services section is the first heading on this page and addresses all aspects of the phone. That includes apps but also systems that are part of the phone's operation

Headings
All of the principle areas of app integration are listed here. So, for example, if an app utilises your Contacts section then, if you tap on the Contacts heading, you will see the related apps within

Battery life
It may be a good idea to disable Location Services for apps and system services which you think can be used without tracking your location to conserve battery life. Go to Settings and then Privacy, tap on Location Services and System Services.

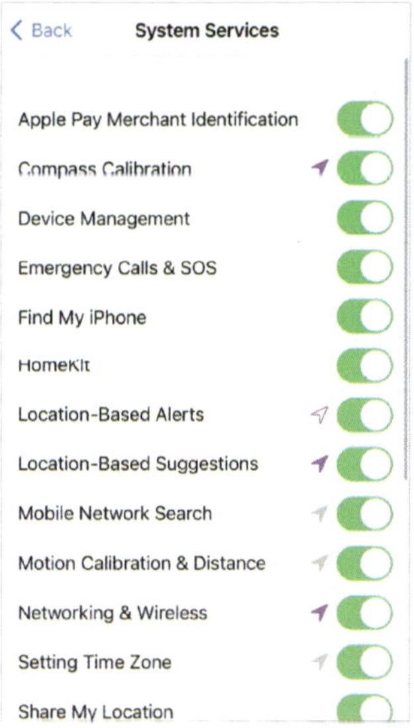

05 System Services

In Location Services, you will find that you can enable or disable this function for System Services such as Cell Network Search, Compass Calibration and much more.

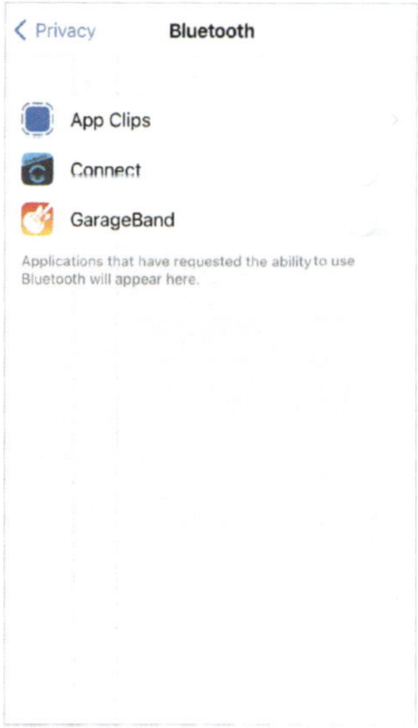

06 Bluetooth privacy

Apps that want to share info via your Bluetooth will appear here. These apps can share data even when you're not using them, so be careful to monitor this section.

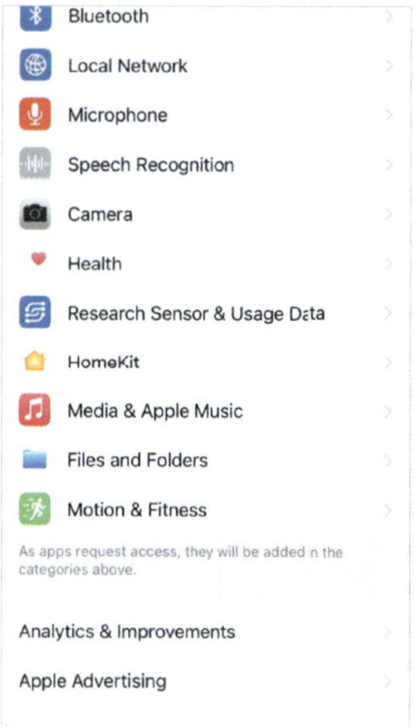

07 Advertising fun

Scroll to the bottom of the Privacy screen to find Apple Advertising. This area can be used to give you more control over advertisers' ability to use tracking methods.

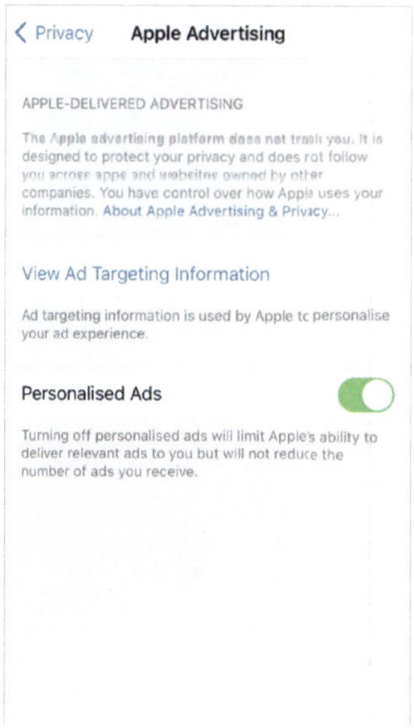

08 Limit tracking

To limit it, tap Apple Advertising to see a new screen featuring just one single switch. Flick to 'On' to prevent apps from using the new advertising identifier to serve you targeted ads.

Enable Dark Mode

Here's how easy it is to change your iPhone's interface

Dark Mode was a feature that iPhone users had wanted for a long time and it seemed like it would never come – but then in 2019 it finally arrived. When enabled, the feature completely changes the appearance of your iPhone's interface, switching most app's background to a darker colour. One benefit of using this feature is that it saves energy usage on your device, meaning the battery will last longer. Another key benefit of using Dark Mode is that it can reduce eye strain and headaches when using your iPhone in low-light situations.

Activating Dark Mode is very easy as it can be found, as with most system changing features, through the Settings app. But this isn't the only method of enabling the feature, as it can also be found through the Control Centre. We will show you in this tutorial how to find and enable Dark Mode in the Settings app.

Settings Turn on Dark Mode

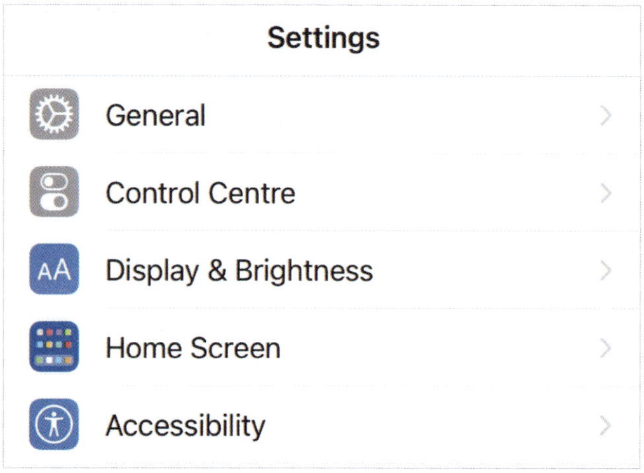

01 Go to Settings

First, open the Settings app by tapping on the icon from the Home screen. It will be in your Dock by default.

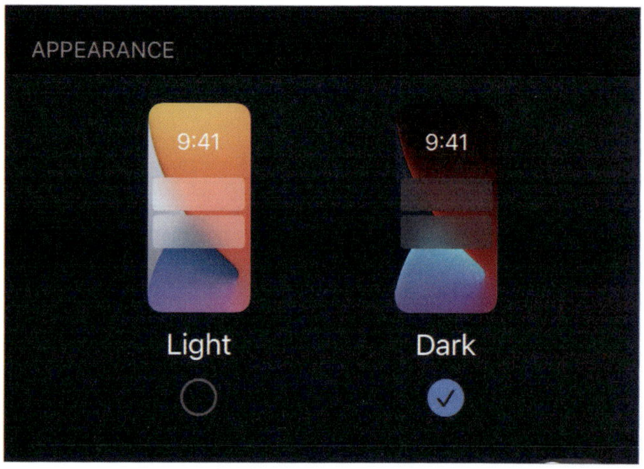

02 Enable Dark Mode

Scroll down to Display & Brightness and tap this option. You will find Dark Mode under Appearance. Tap it and the background will change.

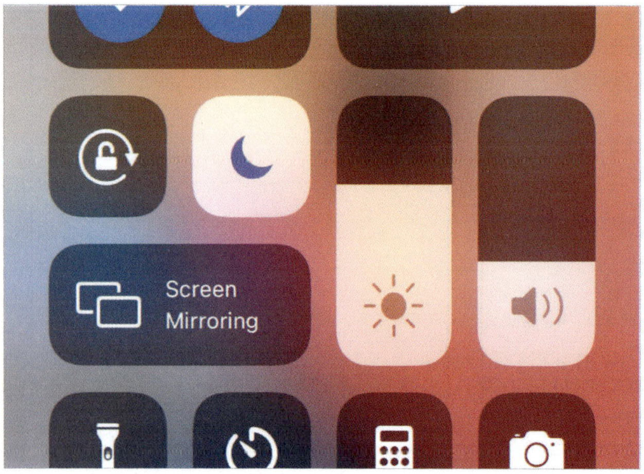

03 Alternative method

You can also enable Dark Mode from Control Centre. Using one finger, slide down the screen from the top-right corner.

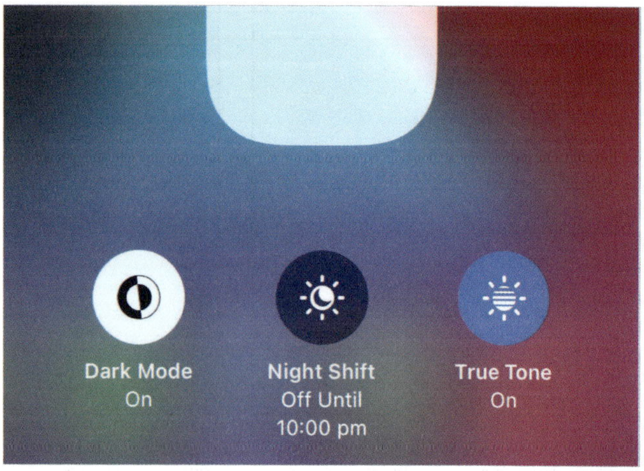

04 Brightness

Tap and hold on the Brightness slider and a series of options will appear. You will find Dark Mode in the bottom-left corner.

Set a Dark Mode Schedule

We explain how to have this feature turn on and off automatically

Dark Mode is available on a range of Apple devices; iPhone, iPad, iPod Touch, and also on Macs. The feature changes the device's interface on a range of native apps so they have a darker background. Despite being able to save a lot of battery life if used a lot, Dark Mode is not something you should have turned on all the time. While having it enabled when in low-light situations can help reduce stress on your eyes, it can have the opposite effect if Dark Mode is enabled during the day. Thankfully Apple has provided an easy-to-use system so you can set when Dark Mode comes on and switches off.

In this tutorial we'll explain how easy it is to create a schedule for Dark Mode. You can select at what time the feature is enabled and also what time it turns itself off.

Settings Create a schedule

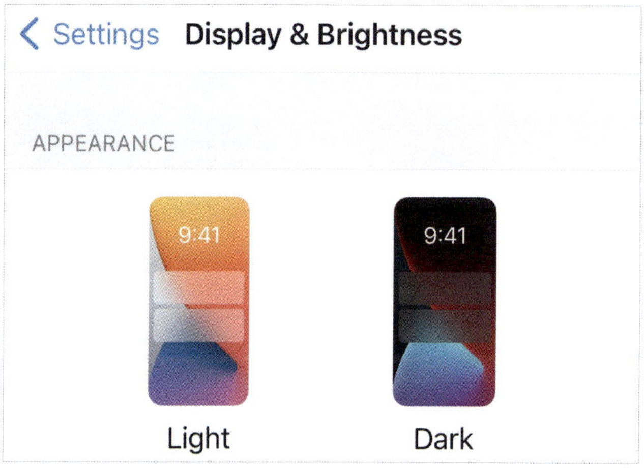

01 Find Dark Mode

Find the Dark Mode settings by going to the Settings app and then selecting Display & Brightness.

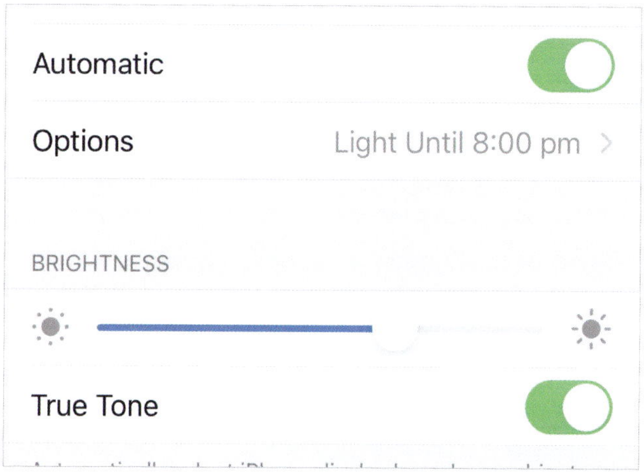

02 Turn on Schedules

Tap the switch next to Automatic so it is on and then select the Options tab to change the schedule.

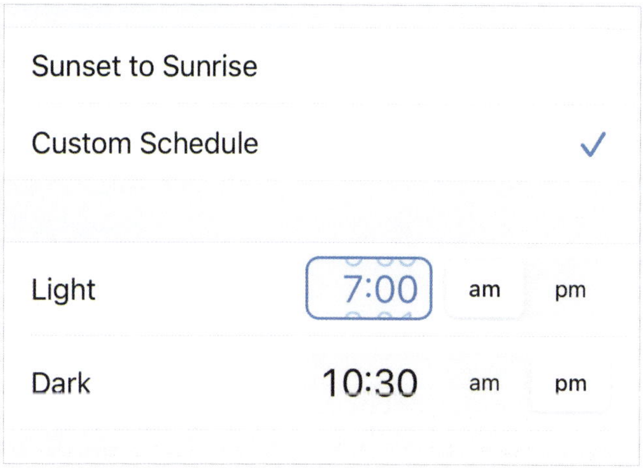

03 Custom Schedule

Tap on the Custom Schedule option and you will see two options; one for when Light Mode is enabled and the other for Dark Mode

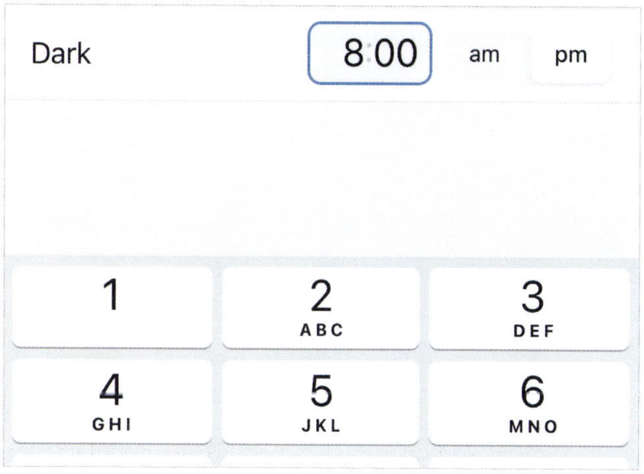

04 Choose your time

Tap on a time, then use the number pad to enter a time and select am or pm. Any changes are automatically saved. Press Back when done.

Use Widgets

Get the latest information at your fingertips

When Steve Jobs demonstrated the iPhone for the first time in 2007, the Lock screen was the first thing he showed. As he swiped across the screen to unlock the phone, the crowd "oohed" and "aahed".

The Lock screen has come a long way since then, and it now offers access to all the information you need at a glance, thanks to widgets. Swipe across the screen and you'll see your widgets displayed here, you can quickly access useful information, like your device usage and the current weather. The News app also gets a large widget, highlighting the biggest news stories and the top two trending features, and tapping on them will open the app.

You'll find third-party apps have widgets, too. It's totally customisable as you can choose what you see and what is hidden, as well as select from a variety of widget styles. Introduced in iOS14, you can drag widgets into your Home Screen so key information can be more accessible. This is the new powered-up widgets function.

Search and find
This Search bar will stay at the top of the screen at all times, even when you scroll. You can use it to search for apps, notes, contacts, and more

Tap to open
The widgets can also work as a shortcut to the apps. Just tap on the widget and the app will open. This doesn't work for every widget though

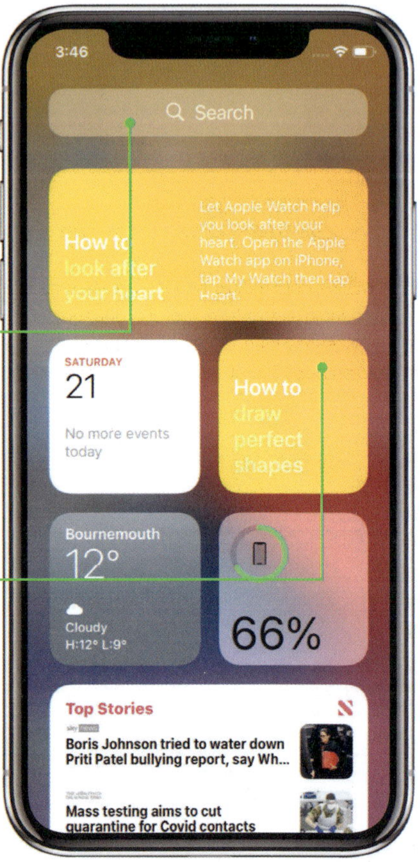

"The Lock screen now offers access to all the information you need at a glance"

Widgets Add and manage widgets

01 Swipe right
To view all of your widgets that are currently available, simply swipe from left to right across Home the screen – just like you used to do when you unlocked your iPhone.

02 Scroll through
By default, your iPhone will have added some widgets to your list that it thinks you might want, such as the weather and top stories from Apple News. Scroll down to see them all.

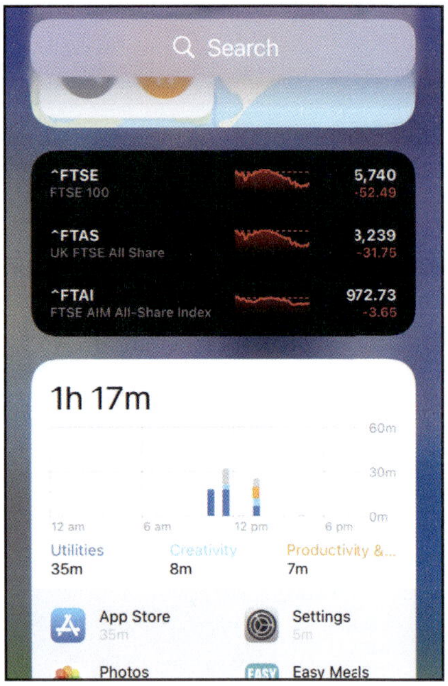

03 Customise the list
Scroll to the bottom of the list; you will see an Edit button. If you want to customise the widgets that appear on this screen, tap the button to open up the editing tools.

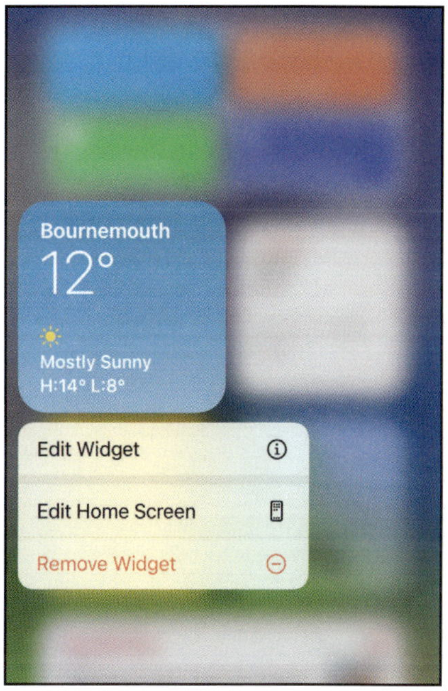

04 Another way to customise

Another method you can use to customise your widgets is to tap and hold on any of the widgets until a menu appears. Select Edit Home Screen and all of the widgets will begin to shake.

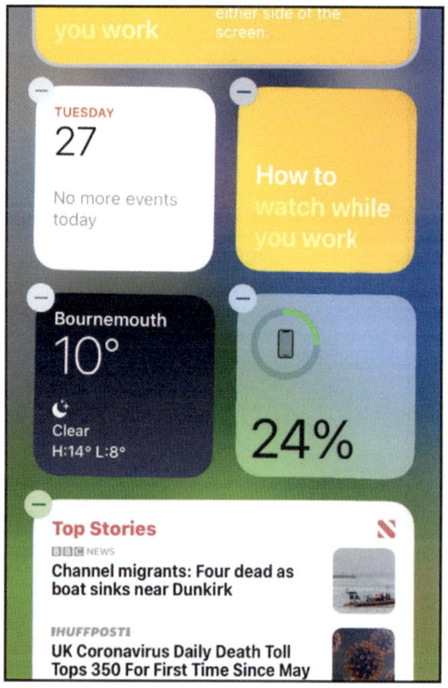

05 Add widgets

Scroll to the top of the Widgets screen and you will see a '+' icon in the top-left corner. Select this and you will see all of the apps that have widgets and can be added to your list.

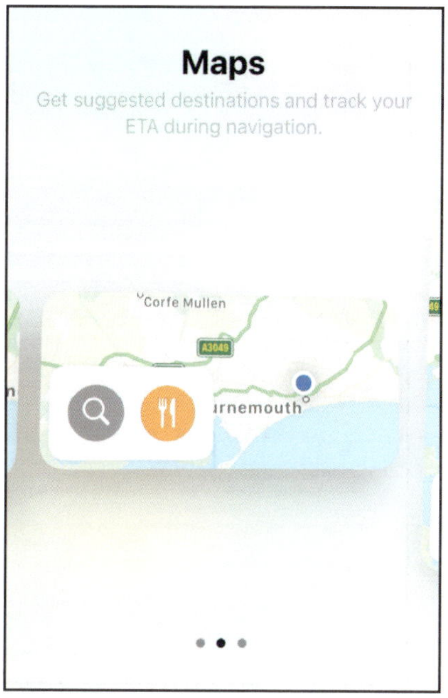

06 Choose a style

Tap an app and you will see all of its widgets. These come in a variety of sizes; from small boxes to long rectangles. When you've found one you like, tap on the Add Widget button.

07 Change the order

You can change the order in which the Widgets appear. Press the Edit button, then tap, hold, and drag the widget around. You can also drag them into the Home screen.

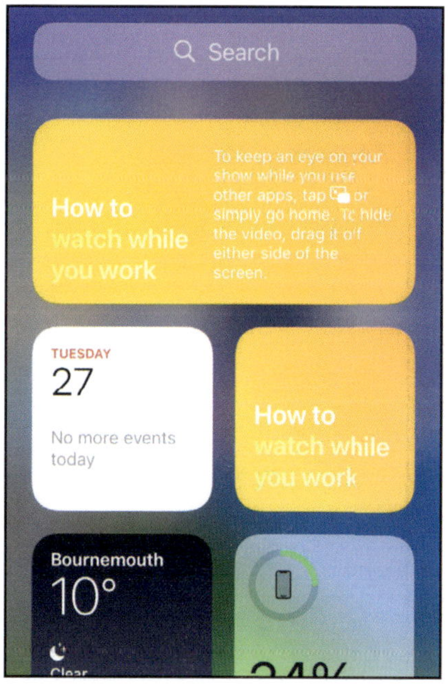

08 Check the list

When you're happy with the order of your widgets, tap the Done button in the top-right. You'll be taken back to the Widget screen so you can check how they look.

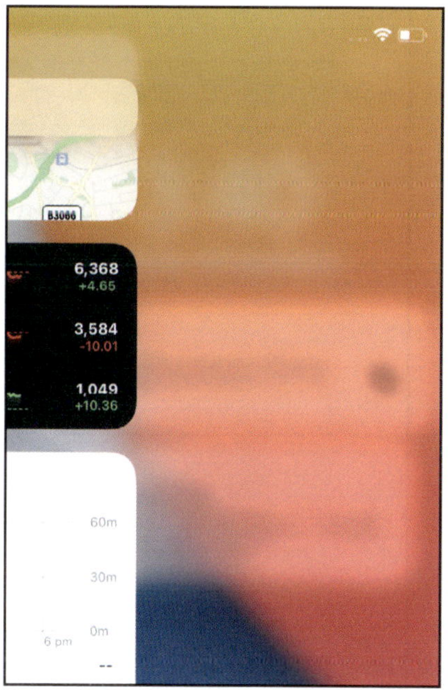

09 Lock screen

You can also quickly view widgets on the Lock screen by simply swiping right just as you would when on the Home screen. To edit widgets you will need to unlock the device first.

Customise your Notifications

Ensure that you never miss a thing in your work or social loop by setting up your own personalised Notification Centre

Your iPhone is rather good at notifying you about updates, messages, events and so on. In the early days of iPhones, the way these messages were conveyed was sometimes quite intrusive, depending on what you were doing, with a message box appearing right in the middle of the screen.

However, the latest iPhones really embrace the concept of notifications, and the system features an excellent suite to allow you to tailor all aspects of how your device goes about communicating messages to you. By going to the Notifications section in Settings, you can set certain apps to notify you via screen alerts whenever something happens – such as an incoming message or email. You can also select between temporary or persistent notifications. You can also check your notifications by swiping down from the top of the screen.

Of course, these settings work the other way too. If you're fed up with being inundated by notifications that you don't need or want to receive, then you can choose to turn off some or all of them. In this tutorial, we'll show you how to keep on top of everything to do with Notifications.

"You can set certain apps to notify you via screen alerts whenever something happens"

Notifications How to set up and use Notifications

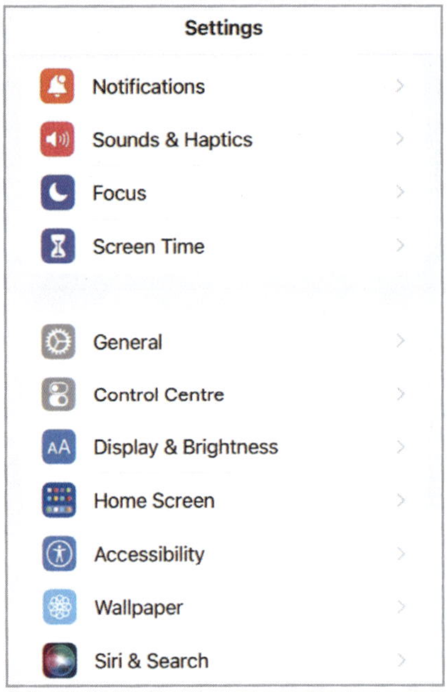

01 Go to Settings

From the Home screen, tap Settings, then choose the Notifications option from the list in front of you. You will then be taken through to the main preferences screen.

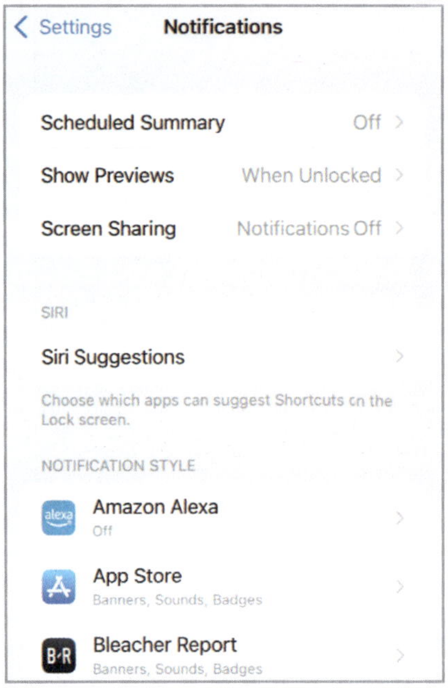

02 Enable/disable notifications

All of the apps that use notifications will be listed here. Most apps will ask whether they can send notifications when you first open them, but you can opt out of them later here.

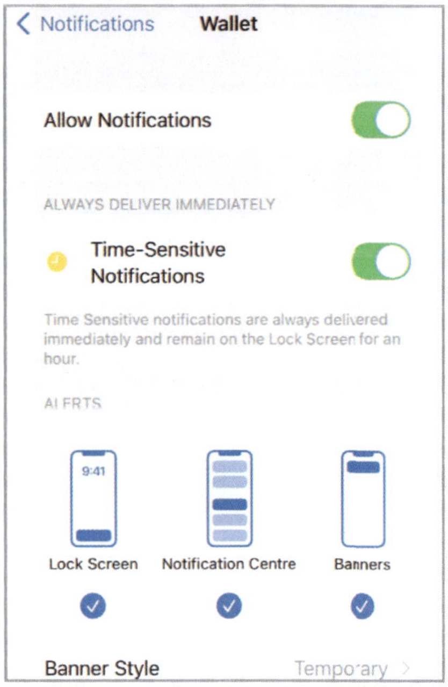

03 Tailor notification options

Tap the arrow next to an app, and you will see all of the options available to you. Choose how many items are displayed and where the alerts appear on screen.

Tweak your notification options

Make sure you are notified in the way you want to be

Badge icons
This slider will determine whether the icon for the app that is notifying you is displayed in the alert. It looks better if they are on

Alert style
You can decide how alerts are conveyed to you; either by the standard style alert, via a non-intrusive banner or choose no notifications at all

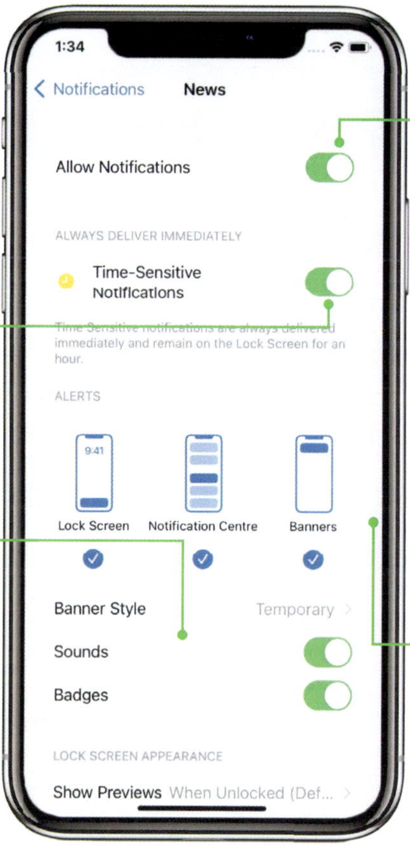

Switch on
To get alerts for particular apps, ensure the Allow Notifications slider is switched on in each app you wish to see featured

Location-based alerts
Some apps can display widgets based on your location. For example you can receive a reminder to call someone when you get to a specific location using the Reminder app. You can turn on these features by going to Privacy> Location Services in Settings and select which apps can use location services.

Notification sounds
Sometimes you may want to see a notification as soon as possible. You can use this slider to determine whether or not your notifications will make noise

04 Access Notification Centre

In order to access your Notifications, swipe down from the top of the screen to open a window of notifications based on the options you have selected that you can scroll through.

05 Dealing with alerts

You can deal with notifications on the fly by swiping it to the left. A set of options will then appear that are specific to the app in question. Simply choose an option from the selection.

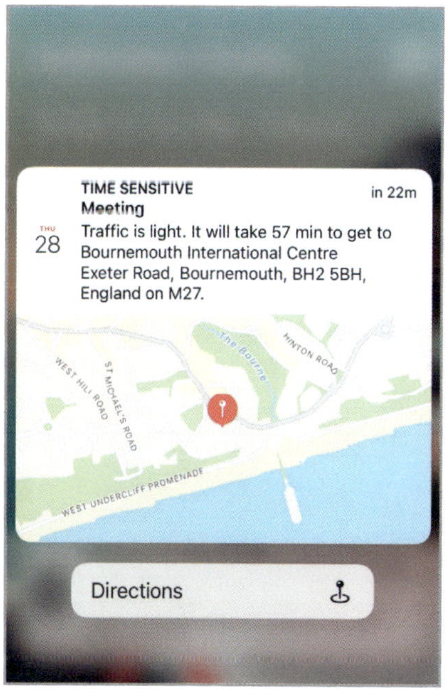

06 3D Touch

You can press lightly on a notification to bring up a menu. Each app's notifications will bring up a different set of options. For example, News will provide a shortcut to the app's top stories.

Change the wallpaper on your iPhone

We show you how to quickly customise the background of your Home and Lock screens

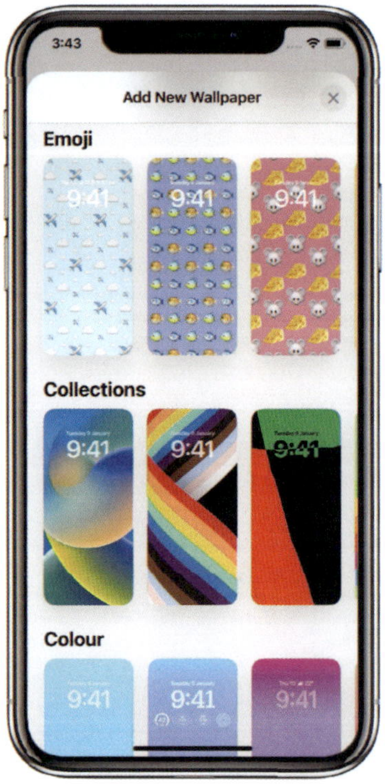

With the iPhone, Apple has given users the ability to change the background of both the Home and the Lock screen. This may seem like a trivial addition to the software set, but it's actually pretty neat. Apple successfully employs a closed system to prevent users from making the system look bad, so we're glad that we get to add a little individuality, and we were also pretty pleased when we found that you can also have different images for the screens.

Making changes to the system is very simple, and proficiency at this simple task should give you the courage to explore the settings further in order to get even more use from your iPhone and make improvements to the way it works for you. Whether it's an image from the iPhone's supplied wallpaper set or a photo from your album – even one that you've saved from the web – following this step-by-step tutorial will enable you to customise your iPhone and have it looking the way you want it to.

"This may seem like a trivial addition to the software set, but it's actually pretty neat"

Settings Change your iPhone's background image

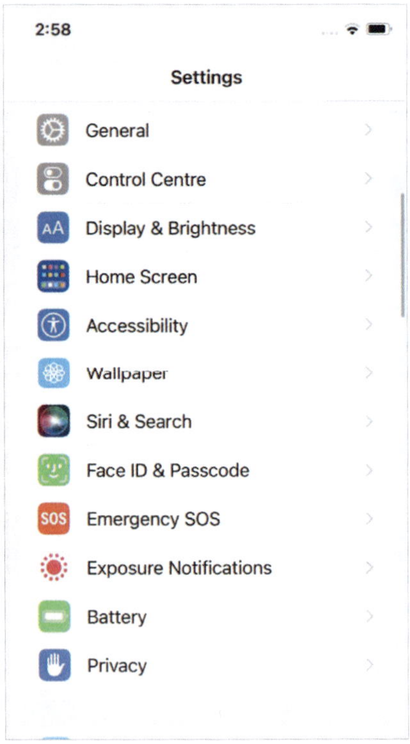

01 Open Settings

Load the Settings app by tapping the Settings icon on the iPhone Home screen, which will take you to the screen shown above. Tap on Wallpaper in the middle of the third section.

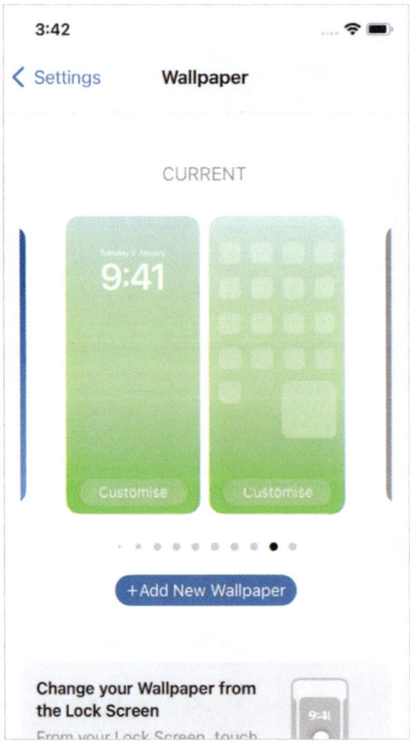

02 Select a screen

These are the backgrounds that you can change: Lock screen on the left and the Home screen on the right. Tap on Choose a New Wallpaper to take you through to the selection screen.

03 View options

You now have several options from which to pick a picture. Choose the album that you wish to pick from, and tap on that album to bring up the contents.

Customise your Home screen

Making the most of your iPhone's screen

Adjustments
Each wallpaper has a series of settings that can be tweaked. You can swipe the image to change its appearance or tap a Widget to customise them.

Cancel it
Apple makes it very easy for you to change your mind and go back to the last action. In this case, just hit the Cancel button

Make your choice
Once you're happy with the image you want as your background, tap on the Add button, and the chosen picture will appear

Great resolution
The iPhone 14 has a Super Retina XDR display screen, offering crystal-clear images. Because of this, you can use photos you've taken on your phone or synced to your device as a wallpaper and they'll still look fantastic.

04 Preview screen

If you select a wallpaper offered by Apple you will be presented with the preview screen. Swipe left or right to switch between the different screens or tap the more icon to see more settings.

05 Edit Widgets

Tap on the Widgets at the top of the screen to change their settings. You can change the font of the time for example. More Widgets can be added for apps such as News, Weather and more.

06 Check it

Once you've tapped an option, you will be taken back to the wallpaper options screen, so just head back to the Home screen to see the change. Use the sleep button to view the Lock screen.

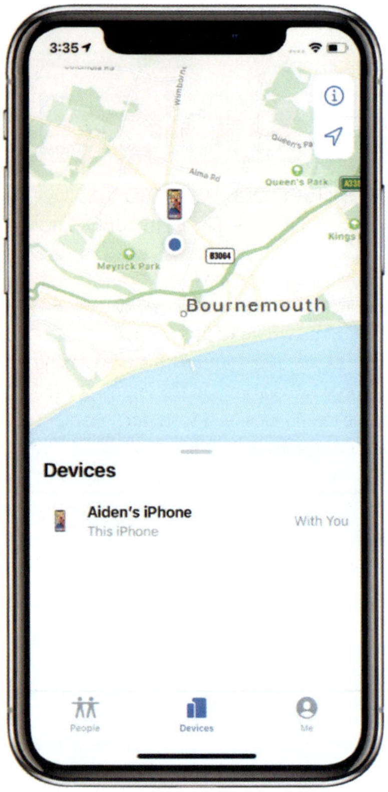

Set up and use Find My iPhone

Find My iPhone is a free service that can be used on any iOS device, and it could save you a lot of money, time and hassle

Find My iPhone is an app and service that you'll hopefully rarely use and is one that will likely find its way to a folder hidden on one of your Home screens. However, this does not mean that it is not vitally important, because if you happen to lose your phone you will certainly be glad that you took the time to set it up properly.

In this tutorial we are going to show you how you can do this and what to do if you lose your phone or – even worse – have it stolen.

It will be worth remembering what to do if you need to take action quickly, but do not put all of your faith in this service. You should still use simple security measures, such as a passcode, to lock the front screen, and of course take all of the usual common sense precautions with a device that is worth a lot of money. There's no doubt that Find My iPhone is very useful, but being careful is always the ultimate solution.

Fingers crossed you will never have to use this app, but here's what to do if you do.

"Find My iPhone is an app and service that you will, hopefully, rarely use, but if you lose your phone you will be glad that you took the time to set it up"

Find My iPhone Be prepared with Find My iPhone

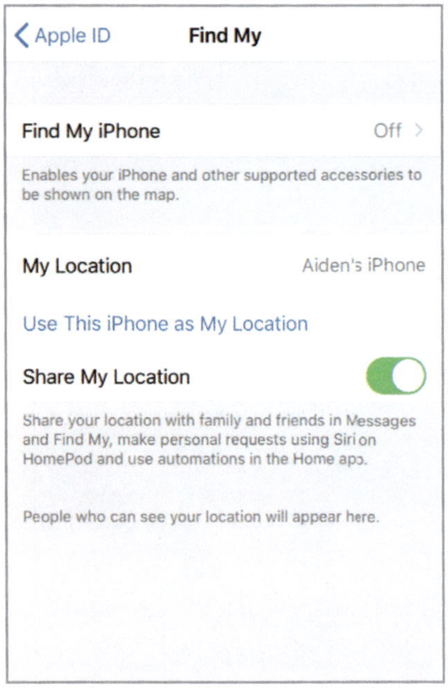

01 Activate the app

Go to the Settings app and tap on your account at the top of the screen, followed by the Find My option. From this menu you can turn on Find My iPhone. Now go to the Find My app.

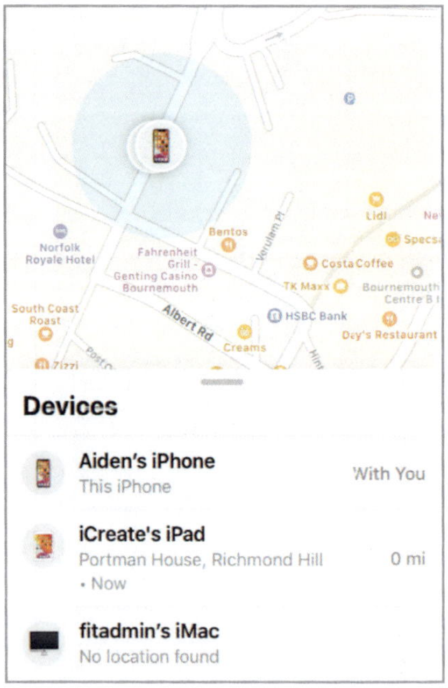

02 Select a device

All of your iOS and macOS devices that are logged into your Apple ID and have had Find My activated will be shown in the Devices tab. Tap on a device to show its location on the map.

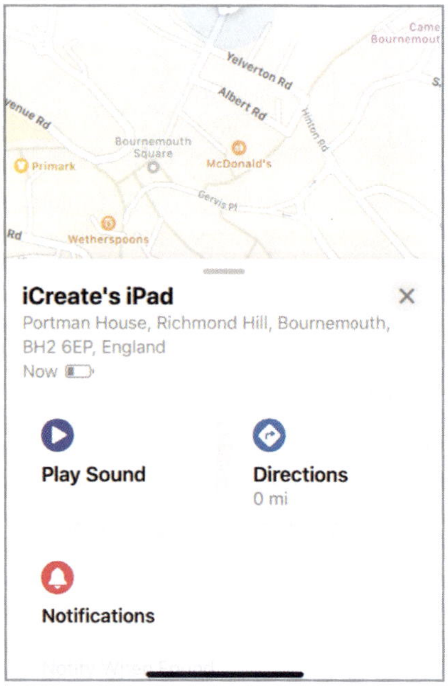

03 Suitable actions

With a device selected, you will see a range of options at the bottom of the screen. Swipe the menu up to see more options. Tap the Directions icon to plot a route to the lost device.

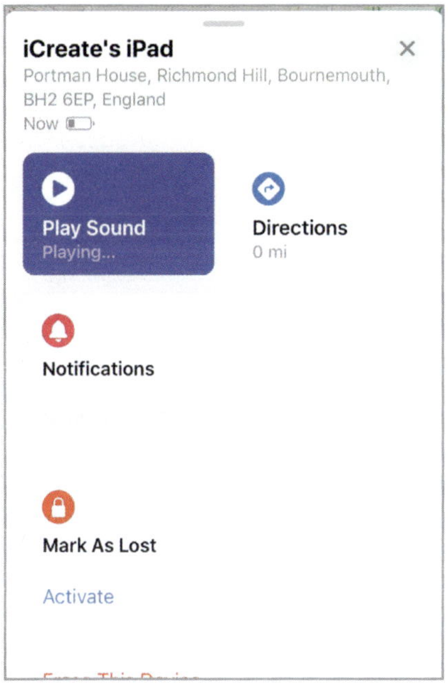

04 A handy feature

The Play Sound feature is found when you tap a device on the map to play an alarm. This is a useful feature if you lose a device locally and have access to another.

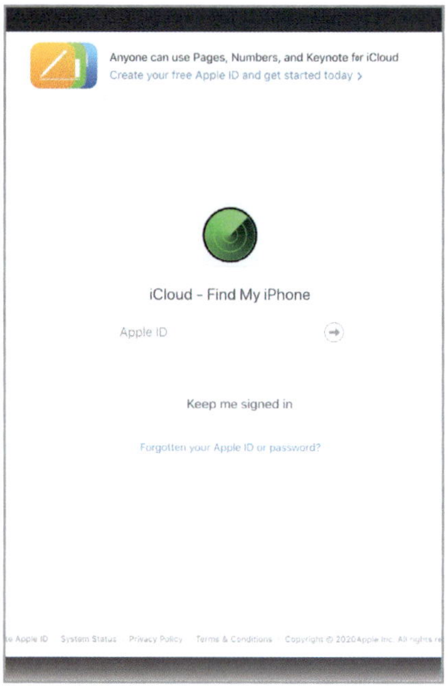

05 Find your phone

If the worst happens, you should first go to **www.icloud.com/find** and enter your Apple ID details. Your phone's location will now be displayed on the screen.

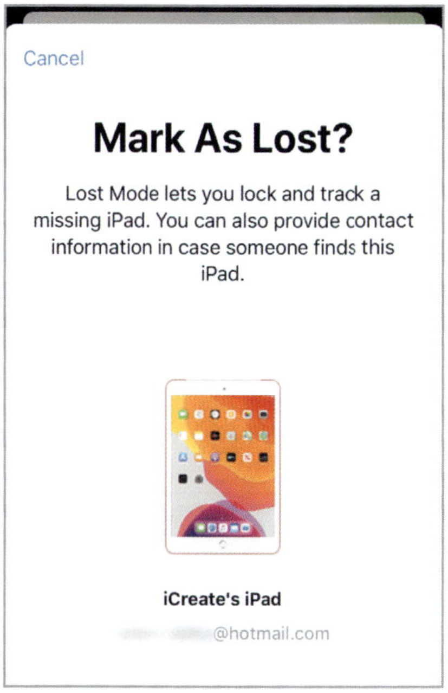

06 Mark As Lost

You can click on the iPhone's location (or the iPad's, if that's what you're after) and select Mark As Lost to remotely lock your device with a passcode to ensure nobody sees what's on it.

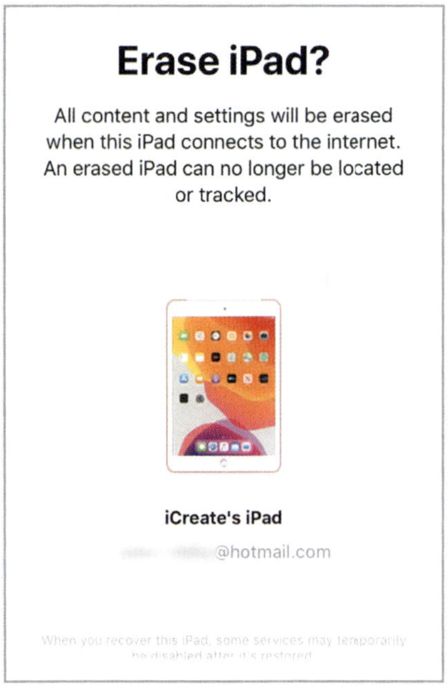

07 If all else fails…

You can take extreme measures. Select Erase This Device, enter your Apple ID and a command will be sent to completely erase the phone. This is a good safety measure against identity theft.

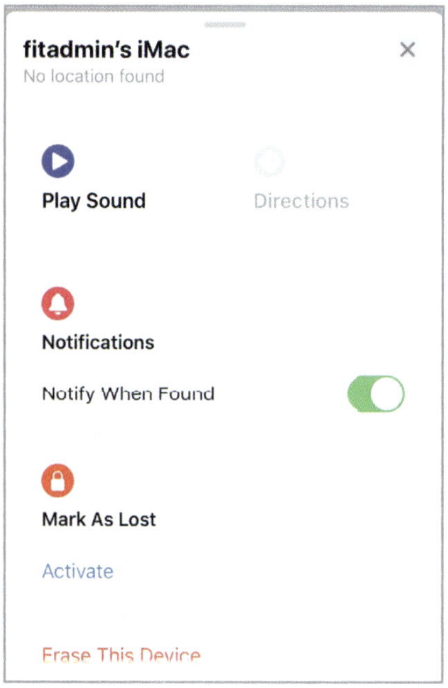

08 Connectivity is needed

The device must be connected to the internet for these measures to work, but you can set up instructions to kick in as soon as it is connected, safeguarding your data.

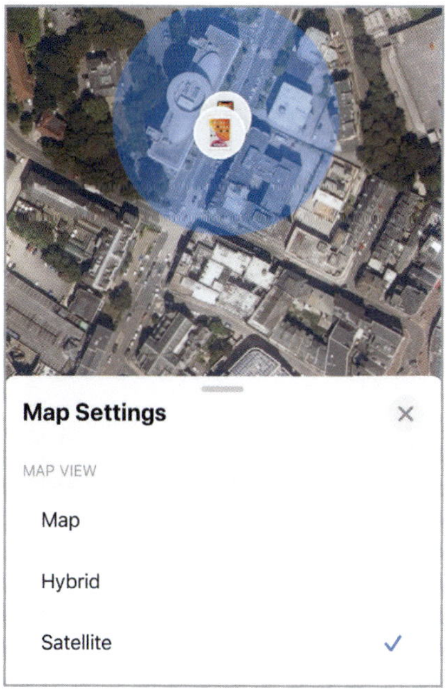

09 Use the satellite

Tap the 'i' icon and select the Satellite view to pinpoint your phone's location. This is particularly useful if reporting the loss to the police who can then use the location as a starting point.

Getting started

66 Master Messages

Everything you need to get up and running with your iPhone

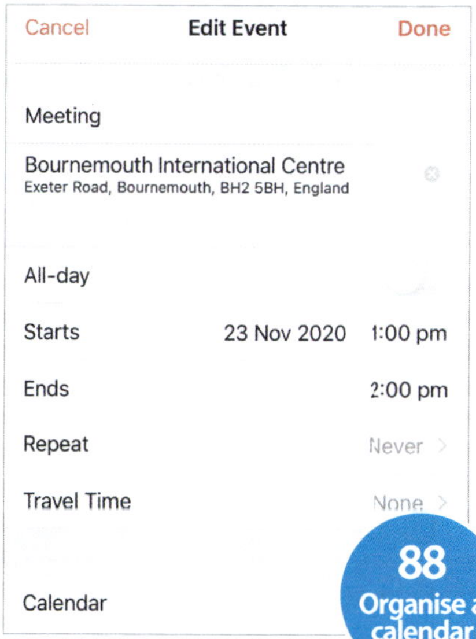

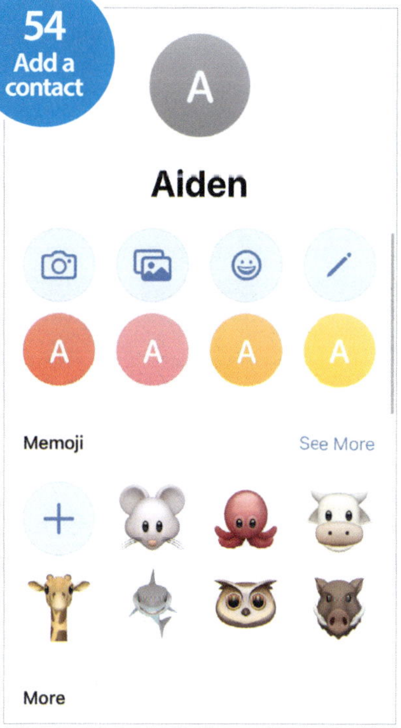

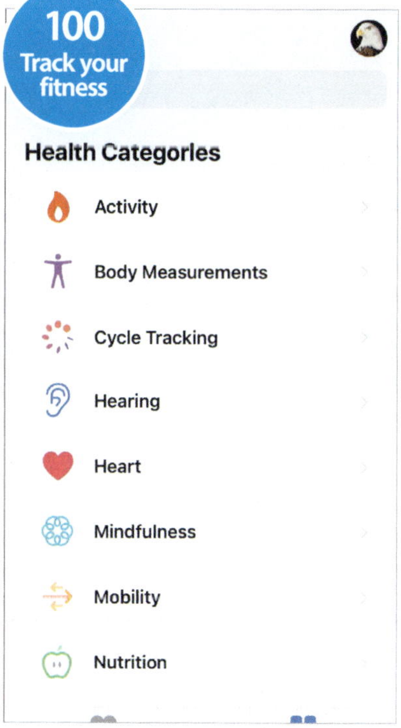

"Once you have your iPhone set up and customised, you can get started by exploring its apps and seeing what it is truly capable of"

Add a contact

Create and update contacts on the iPhone

Now that you've finally got your hands on a shiny new iPhone, one of the first things you're going to want to do is add a contact in the excellent Contacts app. It has easy-to-navigate buttons and a natural feel. If you're new to the whole touchscreen concept, then this process will be a revelation in simplicity. Not only can you add all the pertinent information you need, but there are cool little extras hidden in your iPhone that make the system very slick, intuitive and easy to use.

The Contacts app works when your iPhone is held vertically, and this will give you access to all the options you'll ever need. You can literally feel your way around all of the apps on your iPhone, and the Contacts app is certainly no different. Remember that even if you make a mistake, you can go back and edit anything you like over and over again until you're happy with it. You can also sync contacts from your other iOS and macOS devices using iCloud – so you only ever need create a new contact on any device once.

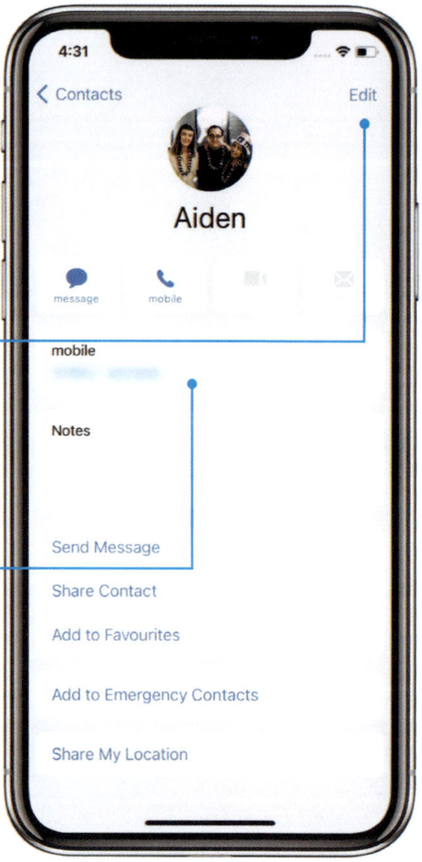

Edit contacts
You can go back to any contact you want and edit their details by tapping on the Edit button at the very top of the page. Hit Done when complete

Contact
Once you've entered a contact, you can talk with them in many ways, including text message and FaceTime

"There are cool extras hidden in your iPhone that make the system slick, intuitive and easy to use"

Contacts Add people to your contacts

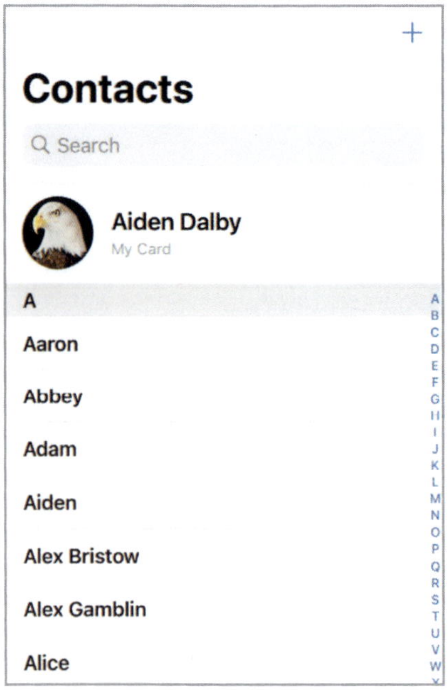

01 Open the app

Open the Contacts app from the Home screen of your iPhone, and then tap the '+' button, which can be found in the top right-hand corner of the screen.

02 Tap and type

After doing this, you can start entering the necessary information. First, tap on the field that you wish to start with. A logical place to start is with the first name.

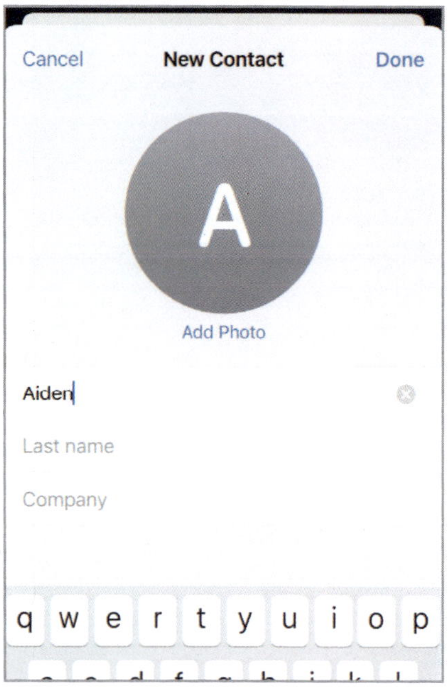

03 Cross it

As you get used to the keyboard, you may find yourself making mistakes. You can use the backspace button to delete characters, or remove everything using the 'x' button.

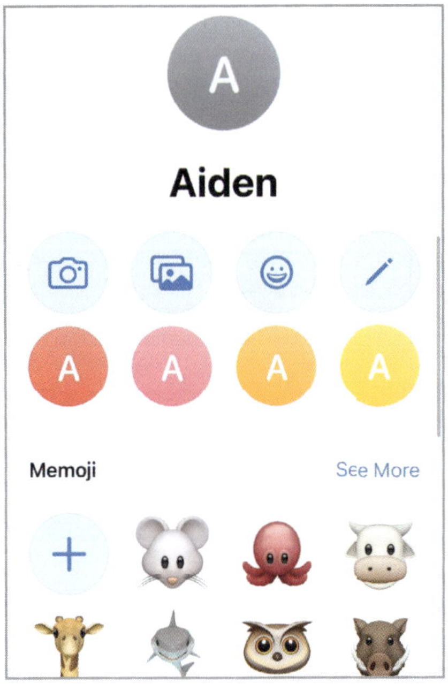

04 Add photo

Tap Add Photo and you will be presented with a range of options. You can choose to either take a new photo, add a Memoji, or add an image from your photo library if there's one you like.

05 Take your pic

Tap the Photos icon and you can now choose from the pictures that you have to hand. If there are more pictures than can fit in the window, then you can scroll up and down to view them all.

06 Move and scale

You are now able to move and scale the picture that you have selected according to your preferences. Pinch or reverse pinch in order to zoom in and out of the picture.

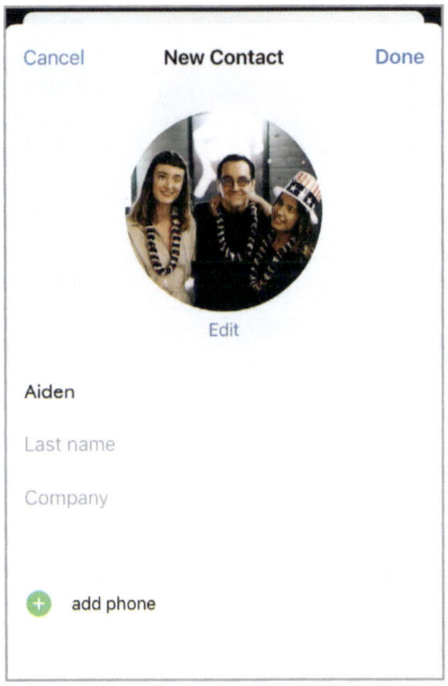

07 Choose or cancel

Once the picture has been arranged how you would like it, tap on the Choose button at the bottom of the window. The picture you have selected will then slide into place.

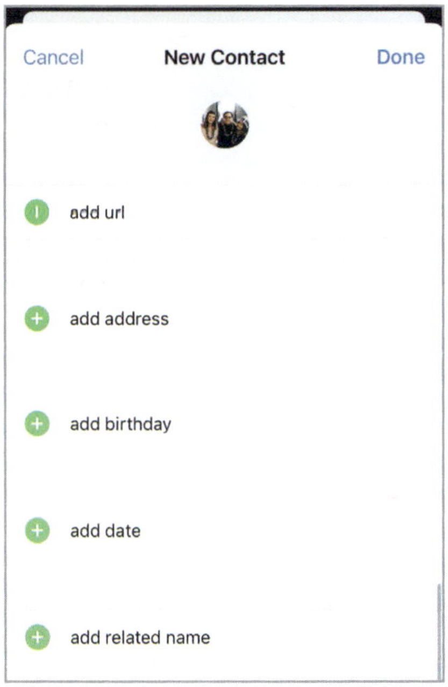

08 Add fields

Go through the rest of the form and add as much or as little information for your contact as you want. Just tap on a field to edit it, and the keyboard will spring up.

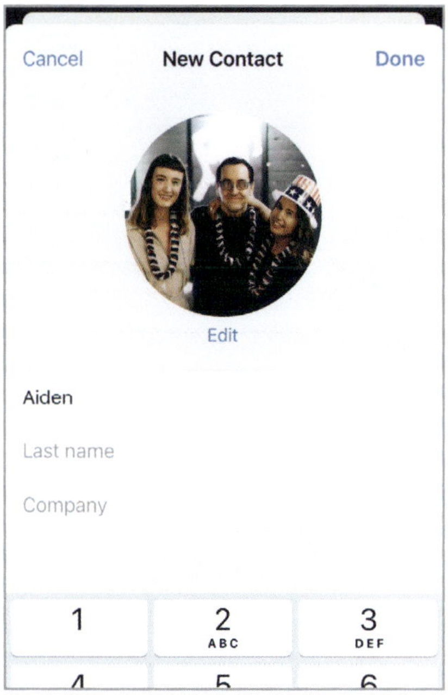

09 Finish up

Once you are completely finished, all you have to do is hit the Done button in the top-right corner of the screen in order to save the changes. Now, proceed to contact your contact!

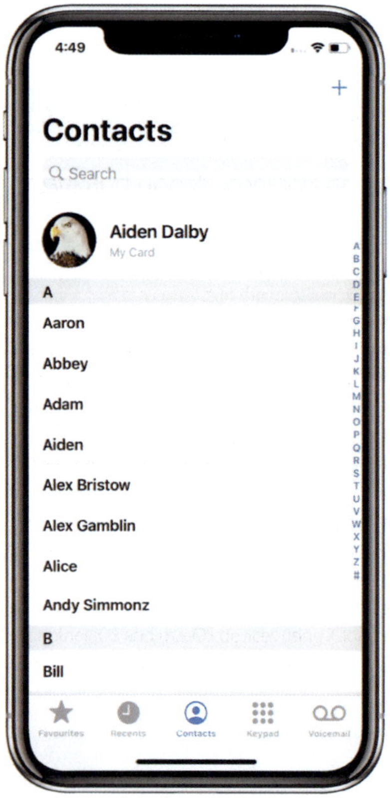

Call a person from your contacts

Simple steps on how to locate contact details and then make a call from your iPhone

Calling a contact from your iPhone is simple, and takes advantage of one of the device's most basic but most essential and commonly used tasks. Once you have added all of your contact details onto your iPhone, it will automatically store them in alphabetical order, making it easy to search and call a specific name and number from your contacts list without having to trawl through endless entries.

You can also use your iPhone's Voice Control feature to locate and call contacts, press and hold the Home button on your iPhone, or request Siri's assistance for completing the task.

Here are some simple, easy-to-follow steps on how to locate your contacts list and then make a call directly by using information. Be careful, when it's this simple, you may find it hard to stop ringing people at all hours of the day!

> "Once you have added all of your contact details onto your iPhone it will store them in alphabetical order"

Phone Search for and call a contact

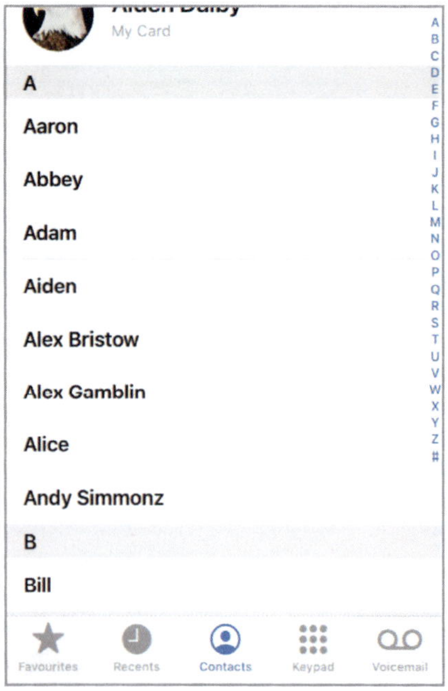

01 Phone menu

Select the Phone icon from the Home screen. This will take you into the phone call out menu. Select the Contacts icon along the bottom bar to bring up all of your contacts in alphabetical order.

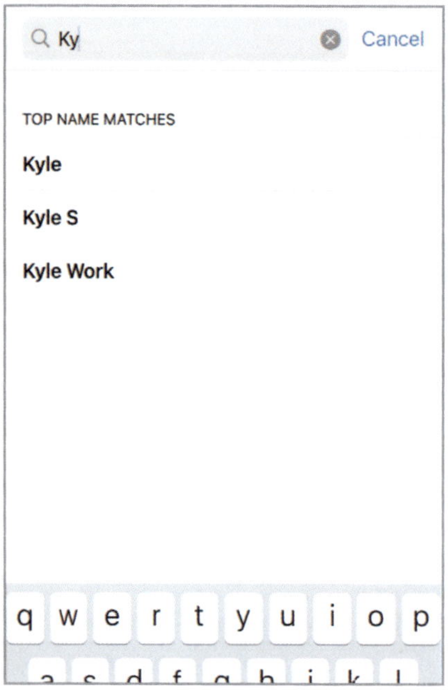

02 Search for a contact

Using the alphabet down the right-hand side of the screen, select the first letter of the last name you wish to select and call. Alternatively you can use the search bar at the top of the page.

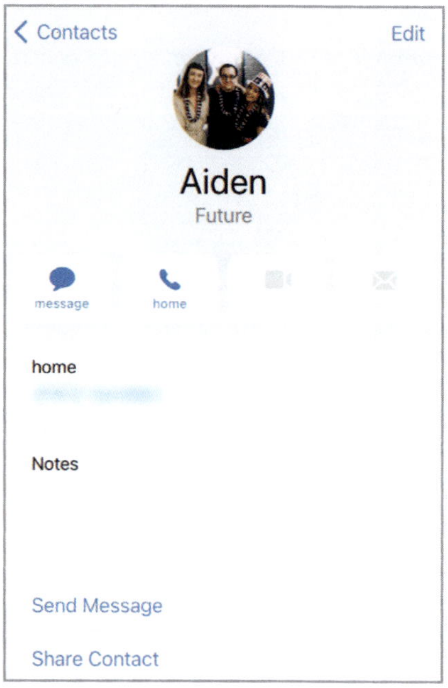

03 Make a call

Once you have located your contact's name tap on it, this will take you onto their information page. Select their phone number (either mobile or home) to make the call.

Make a call using the keypad

Use your iPhone's keypad to dial a phone number that you don't have stored in your Contacts

Using your iPhone's keypad to dial phone numbers and make quick direct calls to people you don't have listed in your contacts is one of the iPhone's main features. This is the same for any mobile phone and works in an almost universal way.

The feature is easy to use; once you have selected the Phone icon from the Home screen you can enter in a number and even add it into your contacts if you wish. The iPhone's intuitive Voice Control feature can also be used to dial a call hands-free when on the go.

Follow along with these three simple steps to access your iPhone's keypad and dial out to make a call. The interface looks a little different if you're running an old OS, but it works in the same way.

"Once you've selected the Phone icon you can enter in a number and even add it into your contacts if you wish"

Phone Dial a call using the keypad

01 Load the keypad

Begin by selecting the Phone icon from the Home screen. Your iPhone should now open the keypad, however you can also select the keypad icon, if you need to, from the bottom bar menu.

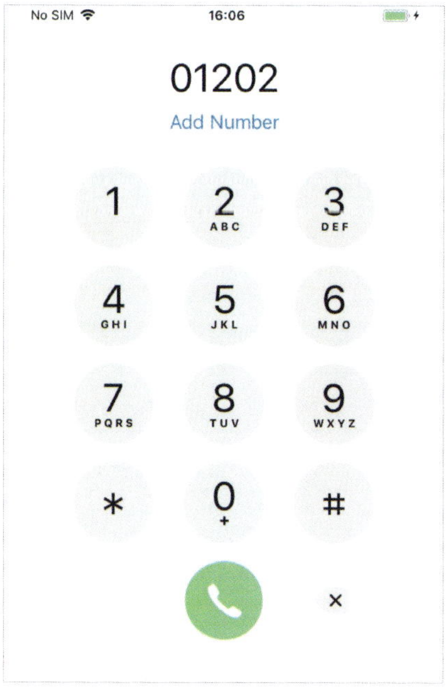

02 Dial numbers

Once the keypad menu is open you can begin typing in the number. You can also add the number to your contacts by clicking the 'plus' icon to the left of the number you have typed in.

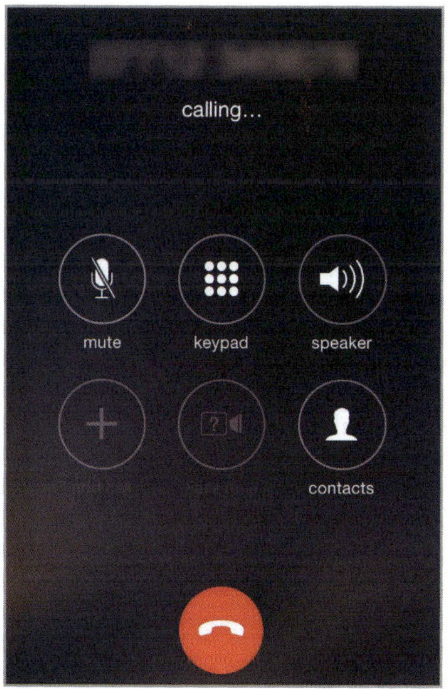

03 Call out

Once you have typed in the required phone number, press the green Call button to begin dialling out. This will take you to the dial out screen. You can end the call by tapping the red button.

Reject incoming calls with a message or reminder

Need to reject a call without upsetting the caller? Send an explanatory text without leaving the call screen

We've all been there. We're in a meeting and our iPhone starts buzzing. It's our spouse/parent/friend but we can't answer the call. However, we really don't want to reject the call and face their wrath later. What do we do? Don't panic, because now you can quickly and easily reassure your caller that you've seen their call and you're not ignoring them.

With iOS 6, Apple introduced a nifty little feature, which enables you to reject a call and respond straight away with a text message – with just one tap. Even better, if you have a habit of forgetting to call people back, you can set a reminder to return the call later.

You can make this feature work for you by setting up your own custom default text message responses such as 'In a meeting' or 'On my way', making it even quicker and easier to reject a call when you can't pick up. So, what are you waiting for? Let's learn how to reject incoming calls in style.

"You can reject a call and respond straight away with a text message – with just one tap"

Phone Reject calls in style

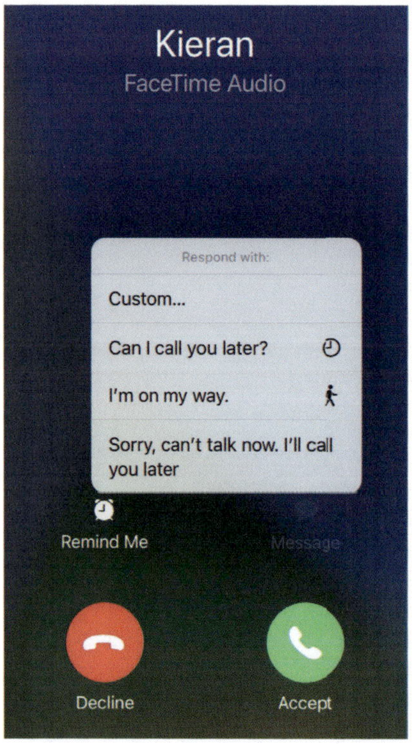

01 Reject a call

When someone is calling you, your screen displays two buttons: Decline and Accept. To see the additional reject call options, tap on the red phone icon and swipe upwards.

02 Reply with a message

To reject someone's call but send them a text, tap the Message button. You will see three default message options, plus an option called Custom to send a text message on the fly.

03 Customise reply options

To customise the default message replies, launch Settings. Scroll down, tap Phone and tap Respond with Text. To create your own message, tap one of the fields and type your message.

Get to know the Phone app

Understand the features which will save you time when dealing with your daily conversations

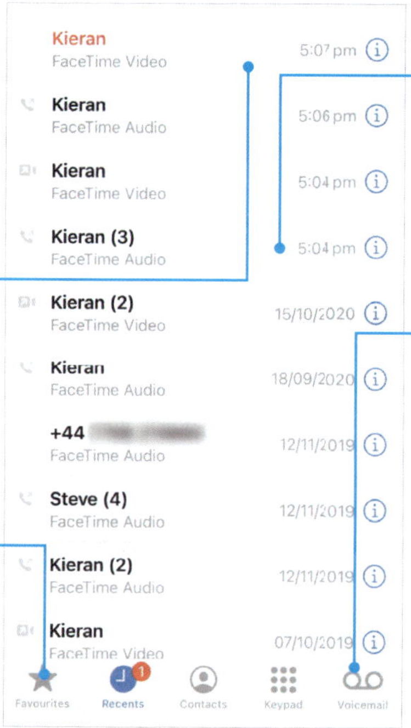

Recent and missed calls

Keep a track of all your recent incoming and outgoing calls by tapping Recents. Tap the missed button at the top of the app interface to see all missed calls

Favourites

Got a few contacts that you speak to regularly? Contact them quickly by assigning them to Favourites. To do this, tap the '+' icon

Contacts

If you don't keep the Contacts app on your Home screen, you can access your contacts quickly and easily by launching the Phone app and tapping the Contacts icon

Voicemail

Quickly and easily access your Voicemail inbox by tapping the Voicemail icon. You'll be dialled through to Voicemail. Some service providers allow you to see your Visual Voicemail inbox

Location reminders

Call return reminders can be location based. Enable this by switching on Location Services for Reminders. When you reject a call and choose Remind Me Later, you'll be reminded when you leave your current location or when you arrive home.

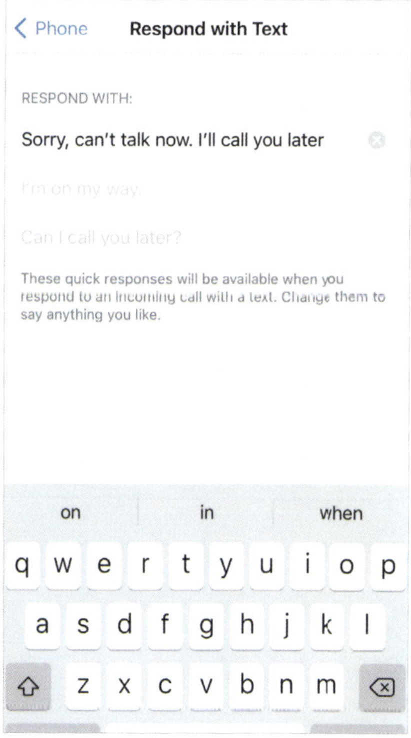

04 Type your reply

Tap and type in each of the fields until you've customised the text message replies to suit your needs. Once you're done, simply tap the Phone arrow button to return to the menu.

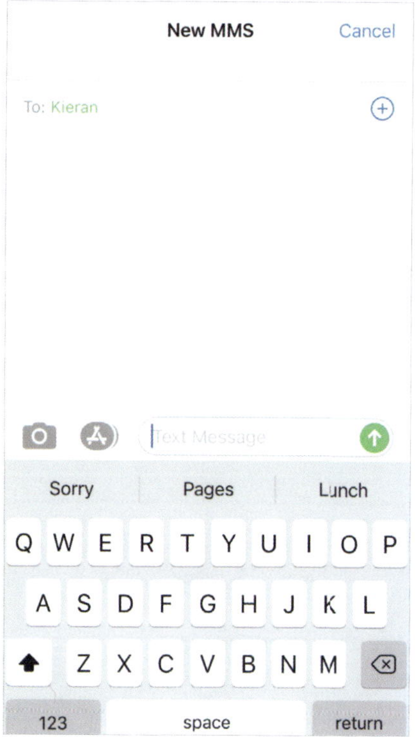

05 Send a message

Tap the Custom button on the incoming call screen to reject the call and send a custom text message. You'll be taken to the Message interface where you can write and send your message.

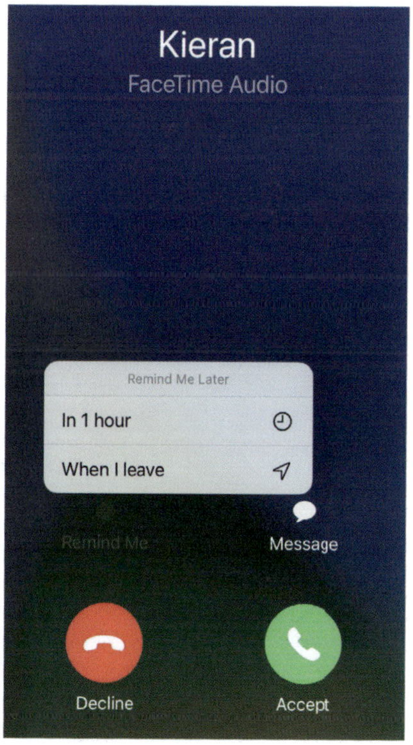

06 Remind me later

Want to be reminded to return the call? You can do so by tapping the Remind Me Later button. You can choose to be reminded one hour later or when you leave your current location.

Understand the Do Not Disturb setting

Take a well-earned break from the pressures of work and life on your iPhone with Do Not Disturb

Do you crave even a few minutes away from the hustle and bustle of daily life? Smartphones enable us to be always online, and always connected to the wider world. The boundaries between our digital, work and personal lives are becoming ever increasingly blurred, meaning that we get fewer opportunities to simply 'switch off' from the distractions that our phones bring.

Apple has come up with a nifty solution to this problem in the form of Do Not Disturb – a feature which, when activated, silences all notifications, alerts, incoming phone calls and messages for a set period of time or until it's deactivated.

Do Not Disturb is a saving grace for anyone who is fed up of being woken in the middle of the night by their phone buzzing or lighting up like a disco ball.

Don't worry though; you won't miss a thing. Your iPhone collects all of the alerts together, while your phone stays dark and silent. Neat, huh? Ready for some peace and quiet?

> "Do Not Disturb silences all notifications, alerts, incoming phone calls and messages for a set period of time or until it's deactivated"

Settings Sleep well with Do Not Disturb

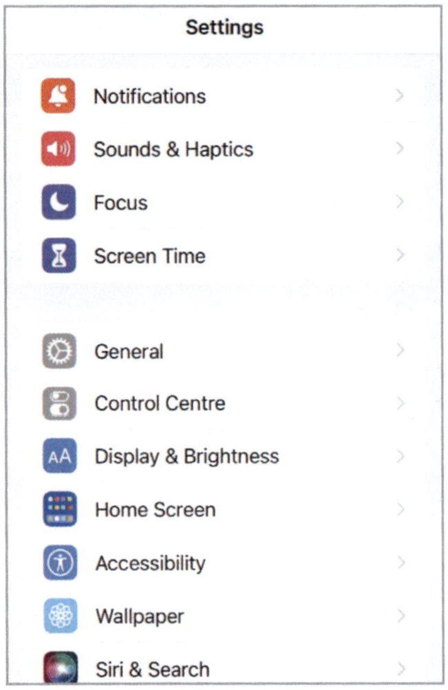

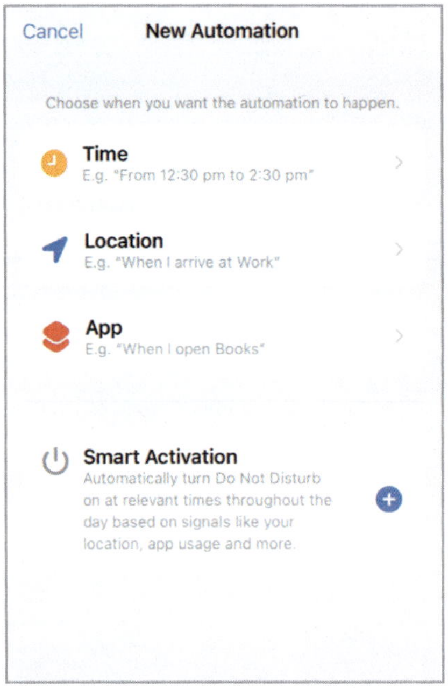

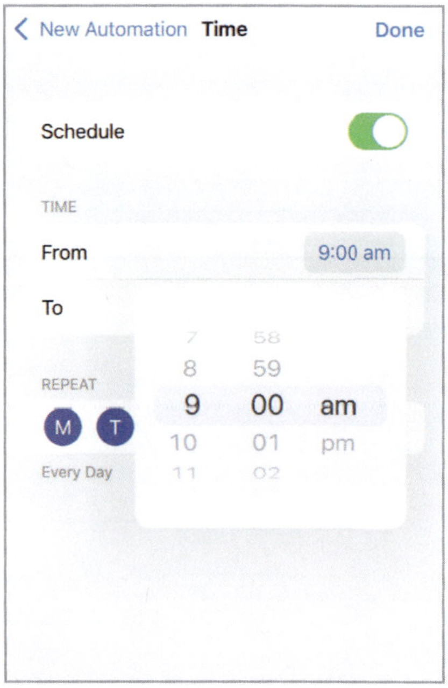

01 Activate Do Not Disturb

Launch the Settings app, tap on the Focus option and then find Do Not Disturb. To instantly enable this mode, move the Manual switch to the On position. Note the crescent icon on the top bar.

02 Scheduled quiet time

To schedule a set time for Do Not Disturb to activate, tap on the 'Add Schedule or Automation' option then tap Time, Location or App. For this tutorial we will be selecting Time.

03 Turn it on

Tap either From or To and a number wheel will appear in the middle of the screen. Enter a time by using your finger to swipe up or down on the time to set it.

Selective silencing

We take a detailed look at how you can customise your incoming call options

Contact Groups
If you would like a specific group of Contacts, such as Family, Work or Favourites, to be able to contact you, whatever the time of day, tap the name of the contact group

Scheduled
If there is a regular time that you'd like to turn on Do Not Disturb (outside of work hours perhaps), then you can use the Scheduled option to automatically enable the feature at certain times

Silence
This handy feature toggles phone call notifications on or off while you're using your phone. For total peace and quiet, select 'Always'

Repeated Calls
Sometimes, if a matter is urgent, you'll get multiple calls in quick succession from the same number. Enabling this feature allows repeated calls to override the Do Not Disturb function

Do Not Disturb your Mac
Did you know that Do Not Disturb is available as a feature on Mac devices too? Open the Notification Centre application, scroll to the top and toggle the Show Alerts and Banners switch to 'Off'.

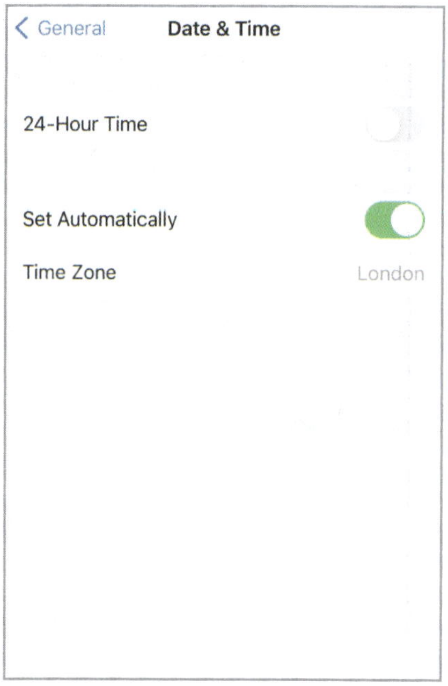

04 24-hour time

For our example we are using a 12-hour clock. You can change between 12 and 24-hour clocks by going into the main Settings app screen and selecting General and then Date & Time.

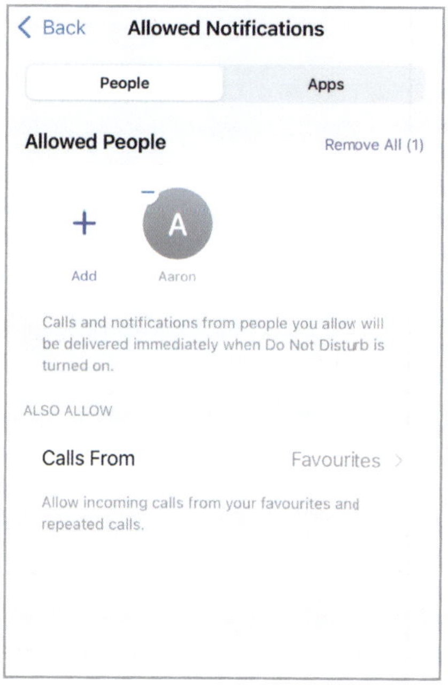

05 Allow Calls From

Tap the Allowed Notifications button and you can select who can contact you during Do Not Disturb times. You can choose from Everyone, No One, and Favourites, or select a Contact Group.

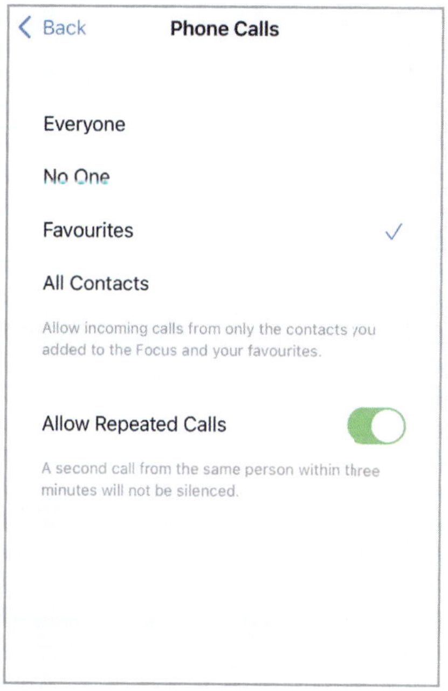

06 Allow Repeated Calls

Sometimes someone may need to get hold of you urgently. Toggle the Allow Repeated Calls switch to 'On' to allow a second call through from the same person within three minutes of their first call.

Chat to friends and family with FaceTime

FaceTime can help you communicate in a way you may not have considered before. It's much more personal than a voice call…

FaceTime is Apple's answer to Skype and it works in a similar way, but is maybe even easier to use. The feature set is kept to a minimum and video calling is the main feature here, but it is the way it works that makes it so useful in day-to-day use. Whether you want to call home when away on a business trip and speak to the kids, or just catch up with a friend in a more personal way, the process to make and receive video calls is the same as making a voice call and only requires a couple of taps to initiate.

Since the introduction of iOS 6, Wi-Fi is no longer required for the calls to connect to a variety of Apple devices – you can now use FaceTime over 4G or 5G. Desktops, laptops, iPads and iPhones are supported and each call is free, although if you're calling using a cellular network, you must keep an eye on your data plan. The use of two cameras on the iPhone means that you can show the world around you as you talk to someone with one tap and that the conversation can become highly interactive if the technology is used to its full potential. This tutorial will help to get you started and, before you know it, FaceTime could become one of your main methods of communicating with others. It really is video calling built for the modern age.

> "FaceTime is Apple's answer to Skype and it works in a similar way, but is maybe even easier to use"

FaceTime Get to know FaceTime

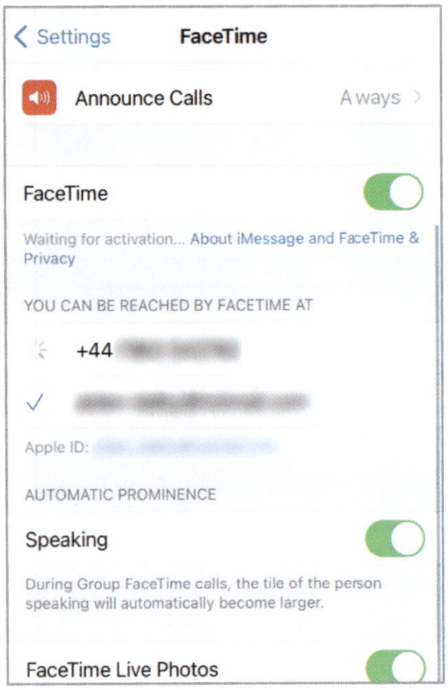

01 Set it up

Go to Settings>FaceTime and make sure that the service is set to 'On', as shown by the green toggle button. You can also define the methods by which you can be contacted.

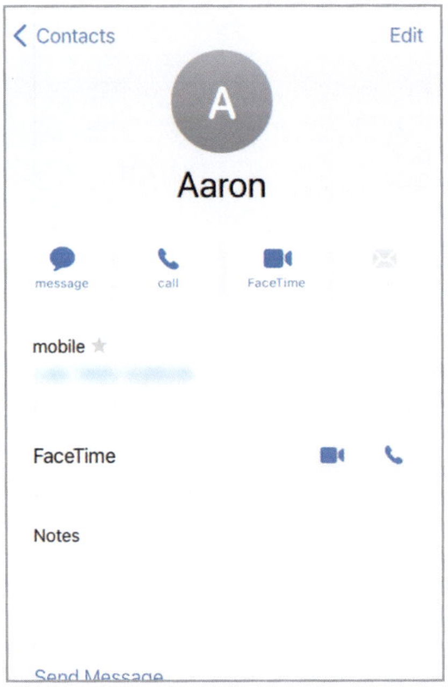

02 Find a contact

Find a contact in your directory who you would like to video call, and look for the FaceTime button that is below the contact's name. When you tap this, a call is initiated immediately.

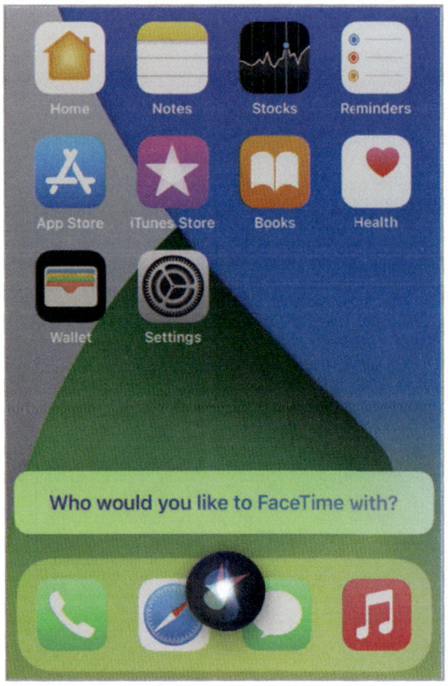

03 Use voice

It's easier than ever to make video calls, as you can also use Siri to initiate a call. Simply say: "FaceTime" followed by the contact's name, and Siri will start the call for you.

Navigate the FaceTime interface

Find your way around everything FaceTime offers

The reverse camera
If you want to show the other person your surroundings, just tap this button and the alternative camera will be activated instantly

Mute call
Tap this button to mute your end of the conversation. Tap it again and it will un-mute you, making you audible once again to the person you're chatting with on the other end of the line

A small window
Your camera output will be shown in a small window at the top right-hand side of the screen. This is useful to check that the other person can see you properly

It's all visual
The person, or people, you are speaking to will always be shown in the main window. They will take centre stage on every call. This showcases the technology really well

FaceTime anywhere
Apple has changed the way FaceTime works over the last few years. FaceTime calls finally work over cellular networks as well as Wi-Fi and this greatly increases the number of uses for the technology. You really will be able to make and receive video calls almost anywhere.

04 The first call

When the call is received by the other party, they will see a message showing the name of the caller (you) and a dialogue explaining that it is a FaceTime call.

05 The magic happens

The face of the recipient will be displayed on the main screen. Your front camera will display your camera output in the small box positioned in the top-right corner.

06 Show your surroundings

If you tap the Flip icon in the bottom-right corner, the cameras will consequently reverse and you can use the back camera to show the person your surroundings.

Manage your iPhone messages

Messages are a quick, simple and easy way to communicate. Here we run through the options to keep yours neat, tidy and current

Messages are a staple diet of any modern day smartphone. They are quick, easy, convenient and almost always shorter than a phone call. The iPhone follows the Apple tradition and adds a little style and panache to the tried-and-tested text messaging experience. The messages are not simply stored in an inbox, they are stored in conversations, exactly as a user would talk to someone face-to-face.

Being colour-coded, users can quickly scan a conversation to see the messages received and sent. The integrated copy, paste and replace commands also mean that users can take even further control of a conversation.

Recent iOS updates have brought a wealth of features to the Messages app. These include animations, stickers, emojis and effects, and each is designed to help you express yourself in new and exciting ways. Of course, you can still send simple text messages without the bells and whistles and this tutorial will guide you through the standard functions of the Messages app to do just that. We will be covering the new features of the app in the coming pages.

> "Recent iOS updates have brought a wealth of new features to the Messages app, including animations, stickers, emojis and effects"

Messages Manage and organise iPhone messages

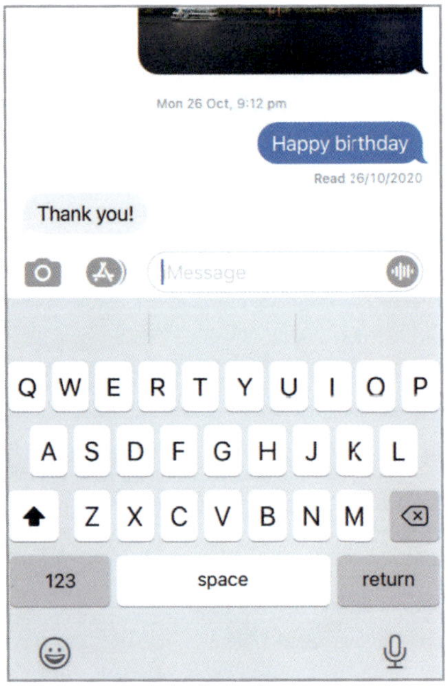

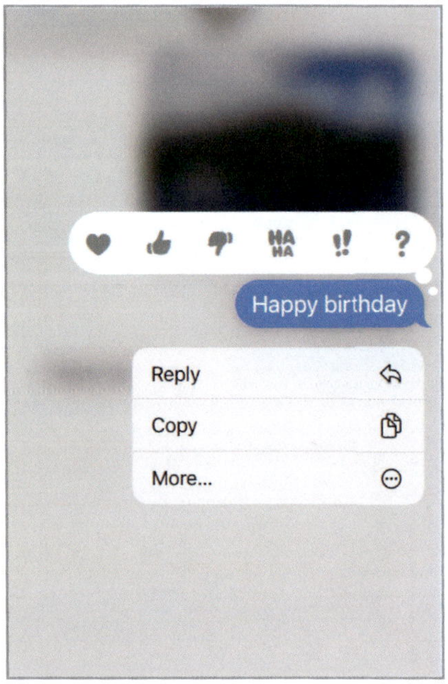

01 View messages

To remove a message, first tap Edit, followed by Select Messages, and then choose the message before tapping Delete. Alternatively, swipe to the left on a message and tap the delete icon.

02 Send a new message

To view a message thread, tap the desired message in the list. To send a message, hit the box at the bottom, type in a message and then simply give the up arrow button a tap.

03 Copy and paste

If you need to copy text into a new message, tap and hold the appropriate speech bubble and tap Copy. Now tap the message box and hold and tap Paste to add.

Make the most of messaging

Get accustomed to the iPhone's text message interface

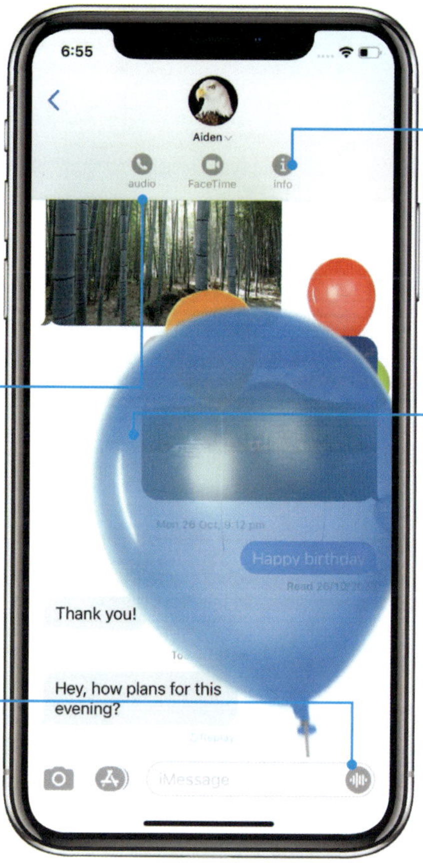

Call and contact

A conversation is related to a specific contact and there are several options for the contact. Tap on the contact image at the top of the screen and icons to Call or FaceTime will appear. You can also share locations and more

Voice messages

If you are sending an iMessage to another iOS user then you will also have the option to record and send an audio message by tapping the mic icon to the right of the text box

Chat details

Tapping on the 'i' icon under a contact's name will bring up the Details screen. Here you can toggle various options, such as muting notifications for a specific chat, as well as viewing all the media that has been sent between chat members in one place

New features

Recent iOS updates introduced many changes to the Messages app, such as animations, stickers and effects. You can find out more about some of these new features overleaf

New emojis

Emojis play a big part in the Messages app. Later in this bookazine you will be able to read all about using and downloading new sets of emoticons, but it is worth knowing that emoticons are now part of the predictive text. If, for example, you start typing 'football', then the option to add a football emoticon will appear above the keyboard.

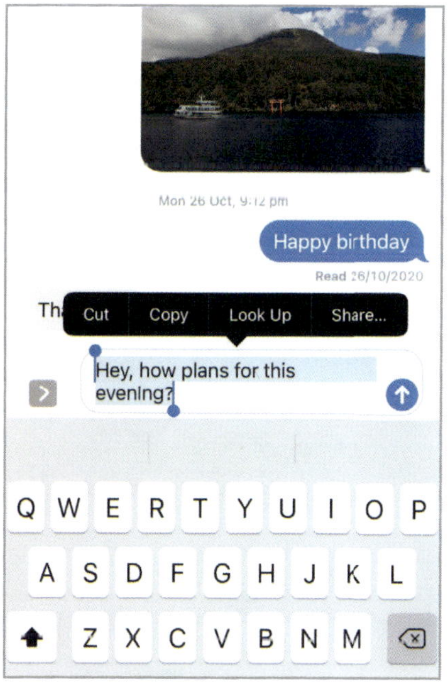

04 Text options

After pasting in text there are options to Select, Select All or Paste, as shown in the shot above. Tap Select to choose a word, then hit Cut to remove, Copy or Paste.

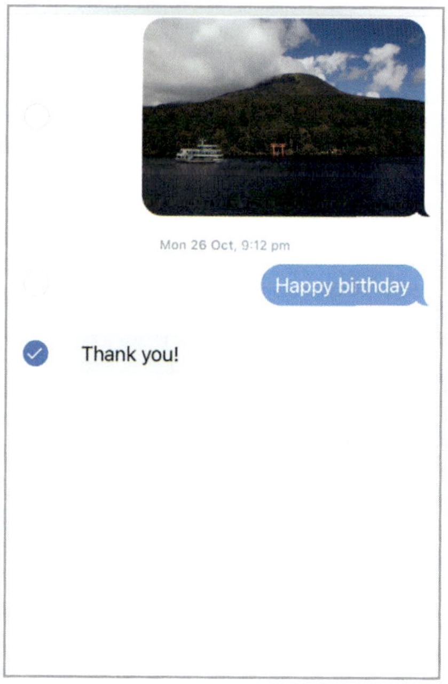

05 Select individual messages

If a message needs to be removed from a conversation, first tap and hold on the message and then choose the More option. Doing so will place a circle next to each message.

06 Delete or forward

Selecting messages activates the Delete and Forward icons at the bottom of the screen. Tap Delete to get rid of selected messages, tap Forward to send messages to a new recipient.

Understand the iMessage features

The Messages app has a plethora of features, including some that are exclusive to newer iPhones

Out of all of your iPhone's bolt-on apps, it is perhaps the Messages app that has the most features to play around with. Rather than just sending bland text messages, you can send animations, a range of stickers, screen effects, Memojis - which are exclusive to newer iPhones – and more. The app has been designed to conveniently accommodate everything, but once you know where everything is you can really start to have fun with your messages. Here we will guide you through the numerous cool features and show you where to access them and how to apply them to your messages. Once mastered, you'll be able to express yourself in new imaginative ways and ensure that the messages that you send to family and friends really brighten up their day.

High impact
Using stickers and Memojis, you can add all manner of special effects to your messages to make them stand out

Features
You can add images, play games and share music in the Messages app – it really is packed to the brim with features

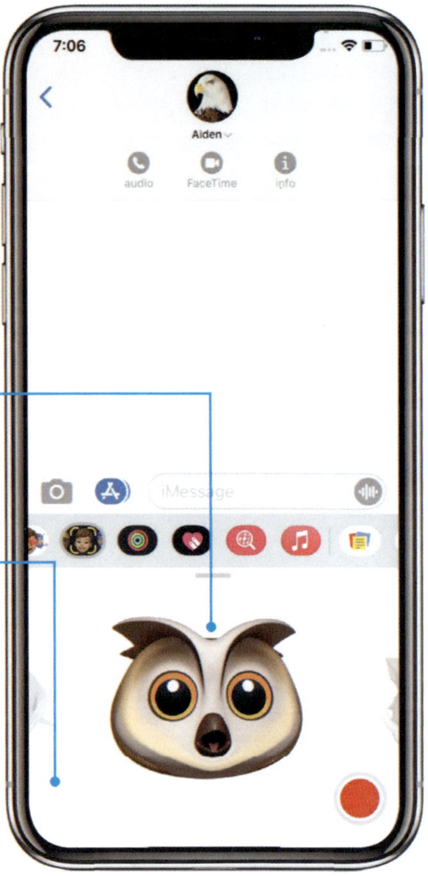

"Express yourself in new imaginative ways and ensure your messages really stand out"

Messages Master the cool messaging features

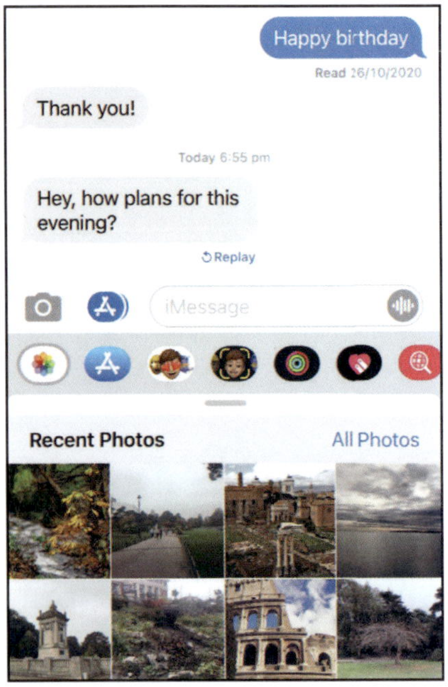

01 Attach a photo
When composing a message, tap on the Photos app icon below the text box. You can either slide the tab to the left to select a recent image or tap All Photos to go though your collection.

02 Take a photo
Another option you have is to tap on the Camera app icon and take a picture using your iPhone's camera. Take a photo and then tap the arrow icon to send it.

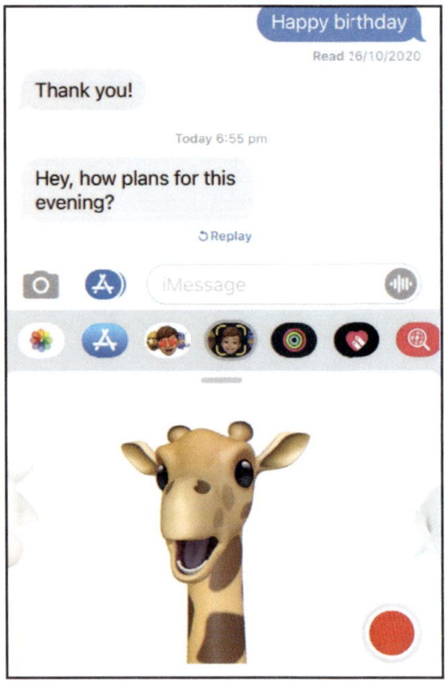

03 Memoji
Tap on the App Store icon, and along the bottom will be a number of apps. Select the animated face icon to record a Memoji; an emoji that tracks your face to mimic expressions.

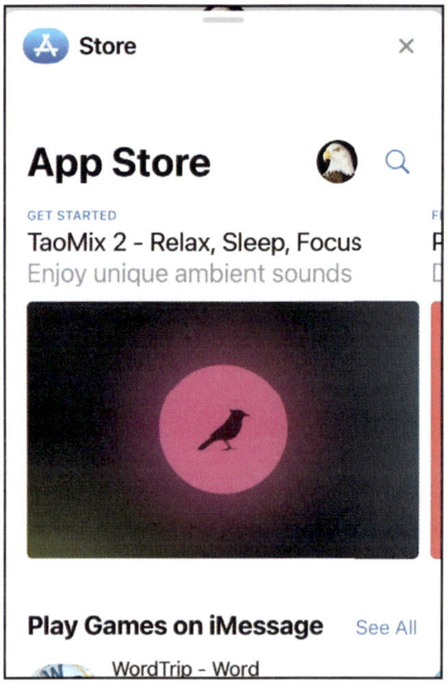

04 Download more content

Tap on the App Store icon again and you will open the store. Here you can download stickers to share via Messages. There are tons to choose from, both paid and free.

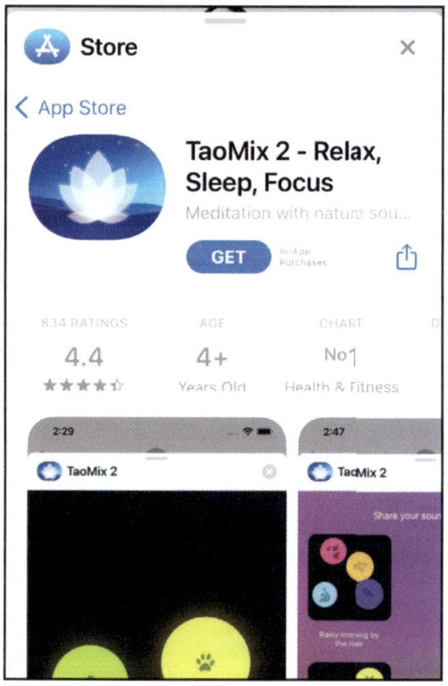

05 Shop for extras

As well as stickers, there are plenty of games to download that you can play with others over the Messages app. Tap on 'Get' then enter your Apple ID password to download the game.

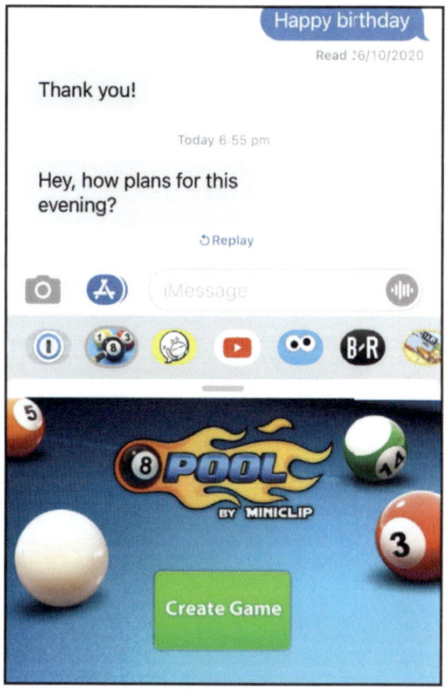

06 Using new content

Once you have downloaded stickers or a game, tap on the Store icon again and you will be able to scroll right through the various icons representing your new content.

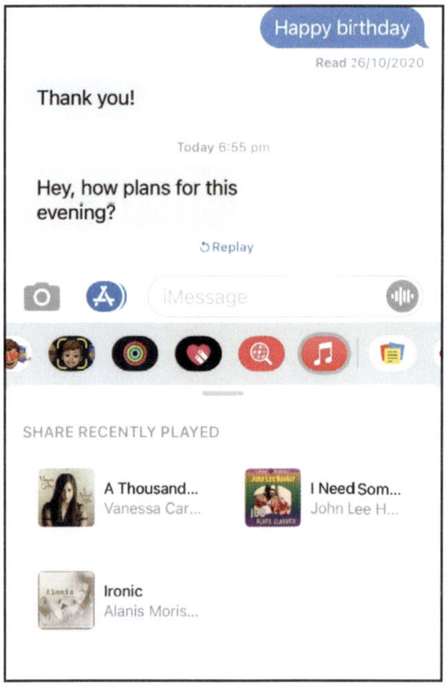

07 Send iTunes recommendations

You can send links to songs you've been listening to. After tapping on the App icon, select the Apple Music icon located along the bottom of the screen and your recently played songs will be listed.

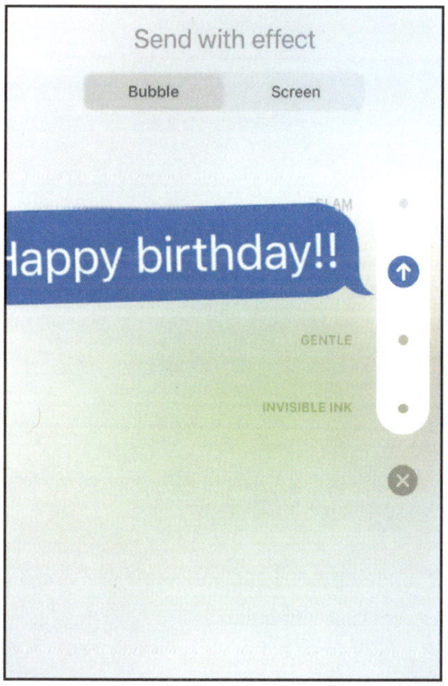

08 Send bubble effects

When sending a message, once you have written your text, press and hold on the up arrow and you will be able to choose a bubble effect from the menu. Tap to choose one.

09 Send screen effects

You can also send screen effects by tapping on the 'Screen' tab at the top of the previous screen. There are several to choose from, so swipe left or right to cycle through them all.

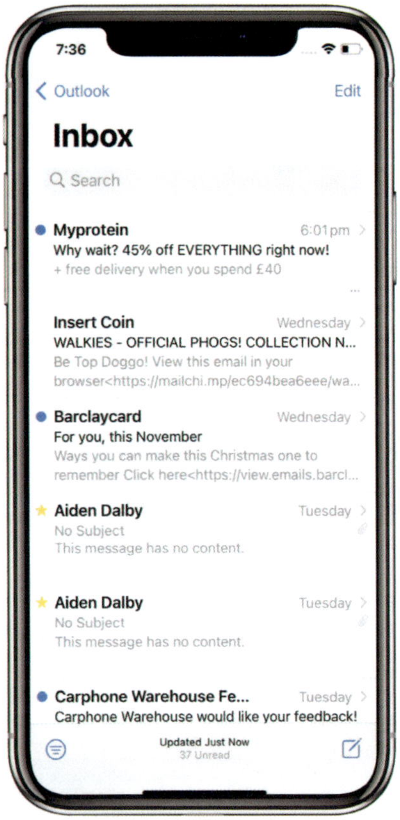

Access emails on your iPhone

Having your emails with you wherever you go makes Mail invaluable. This handy tutorial will help you set up your own personal account

Using email is one of the most essential ways of being able to stay in touch with friends and family, as well as being a pretty vital tool in the business world. The ability to take your messages with you wherever you go on your iPhone makes it even more essential, and means you're not limited to responding whenever you happen to be able to log in to your main email account. Important business can be taken care of as soon as it arrives, and you never need to worry about what could be waiting for you when you finally get back to your desktop.

This step-by-step tutorial will not only show you how to set up a new or existing email account for use on the iPhone and beyond, but will also take you through the fundamental aspects of reading and sending email. Once set up, you'll be able to use existing accounts at will, whether your business or personal accounts, or freebies from the likes of Google or Yahoo!.

You'll be able to reply to and forward mail that you receive and, most importantly, ensure that you stay in touch with friends and loved ones, no matter where they – or you – might be. Basically, you will never look at stodgy old email in the same way again, so read on to learn how to master email on the move and free yourself from your desk at work or home.

> "Important business can be taken care of, and you don't need to worry about what could be waiting for you"

Mail View your emails anywhere

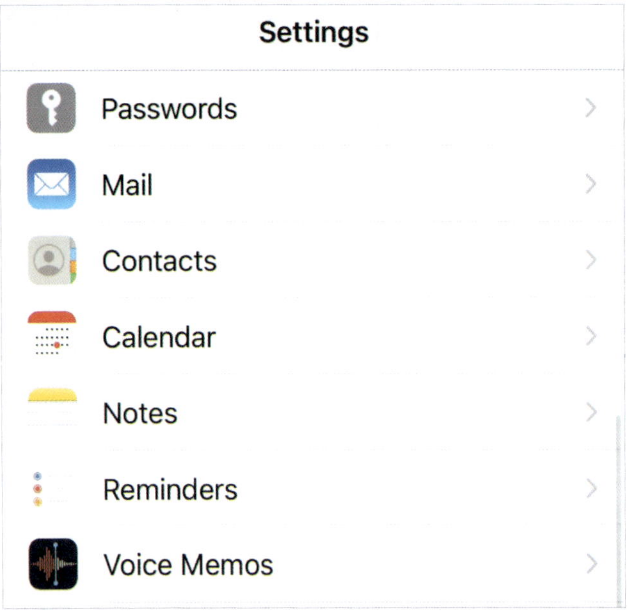

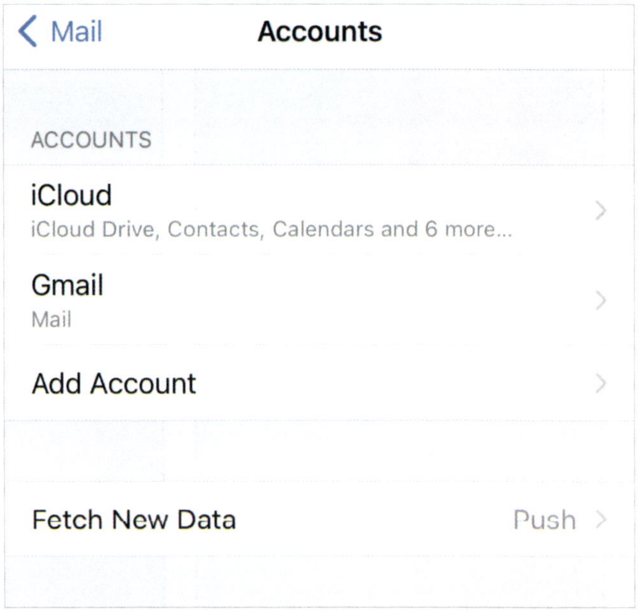

01 Set things up

In order to set up an email account, you will need to first enter the Settings app on your iPhone. Look at the icons on your iPhone's Home screen until you find one with a large cog and two smaller cogs. Tap on it to continue to the Settings menu.

02 Find your mail

Upon entering Settings, scroll down until you see a section called Mail option and then tap on it. Tap on Account which will list all accounts currently on the device. To add a new one, simply tap on the Add Account option that sits beneath the existing accounts, if any.

Viewing mail

How to navigate your messages in Mail

Send email

To send mail, tap on the button in the bottom right-hand corner of the interface. Fill out the address, add a subject and compose your email message. Simply hit Send when you're happy with your note

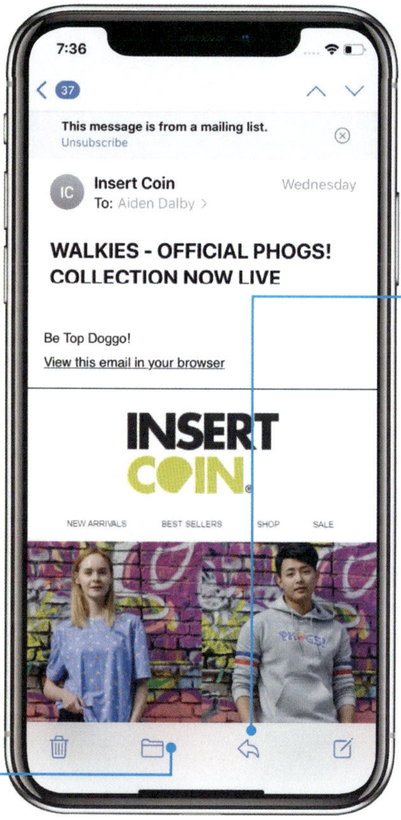

Reply and forward

If you need to reply to an email, hit the arrow icon near the bottom of the screen. You can then reply to the sender or forward the message on

Adding multiple signatures

Like many, you probably have a number of different email accounts, each with their own purposes and uses. Bringing them all into your iPhone is an efficient way of keeping track of them all, but keeping them separate when you respond to or send an email requires new signatures for each. You can easily edit your default signature and create bespoke signatures for each account. Head to the Mail section in the Settings app and select the Signature option to personalise each of your accounts.

Move items

Want to organise your emails? Simply tap the folder icon located at the bottom of the screen. You can then send an email to a variety of different folders

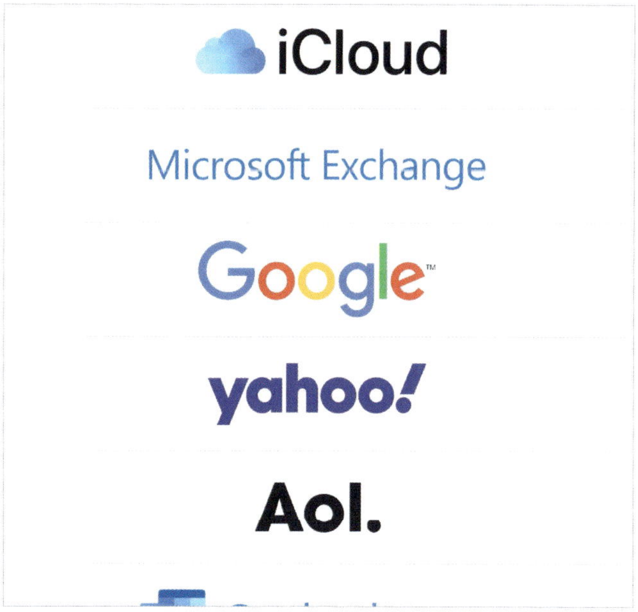

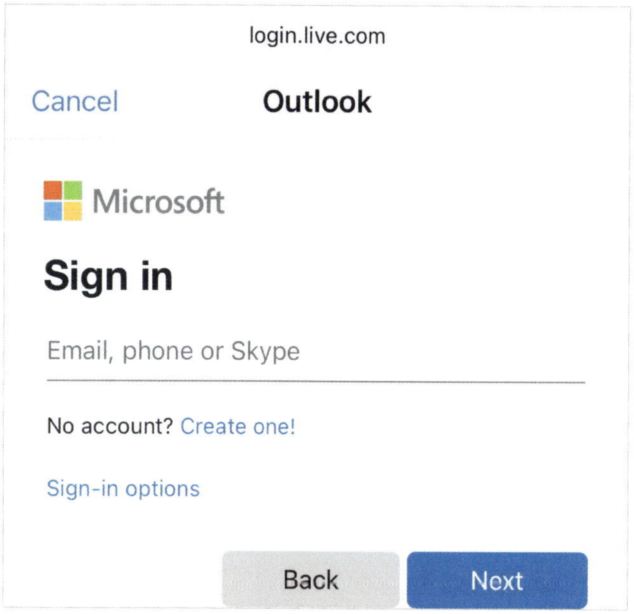

03 Make your choice

You'll now be presented with eight different account options. They are iCloud, Microsoft Exchange, Gmail, Yahoo!, AOL, Windows Live Hotmail, and Other. Whether you want to create a new account or add an existing one, the process is similar.

04 You've got mail!

After choosing your account you'll be presented with a screen which will differ depending on the type of account you're signing into. Once you've logged in, go to the Home screen and tap the Mail app. Congratulations, you've got mail!

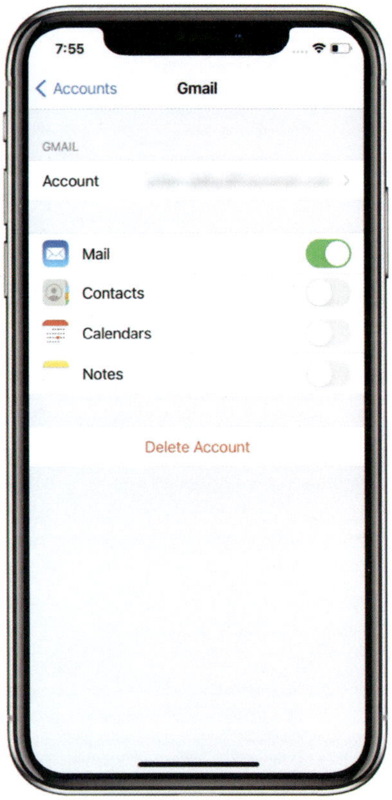

Manage personal and business accounts in Mail

Customise how you interact with your different email accounts, so that Mail suits the way you work

Mail on iOS offers a way to be at your desk even when you're not physically there. The native Mail app makes it effortless to hook up to your various email accounts, regardless of provider, so you never have to be without your correspondence. Whether its a work account or that of your significant other, you'll be able to manage all your messages on the go.

Of course, there is a level of management that comes with having several email accounts in one place, but fortunately Mail has been designed to offer plenty of that. This tutorial will take you through some of the ways you can customise accounts and manage the different content that comes through, including how it appears. It's your email after all; it should appear how you want it to. You might want to prioritise which emails automatically come through, set up a professional signature for your work email account, or sync your contacts, calendars and reminders. With that in mind, let's take a look at how you can personalise things.

"The native Mail app makes it effortless to hook up to your various email accounts, regardless of provider, so you never have to be without your correspondence"

Mail Take control of your email

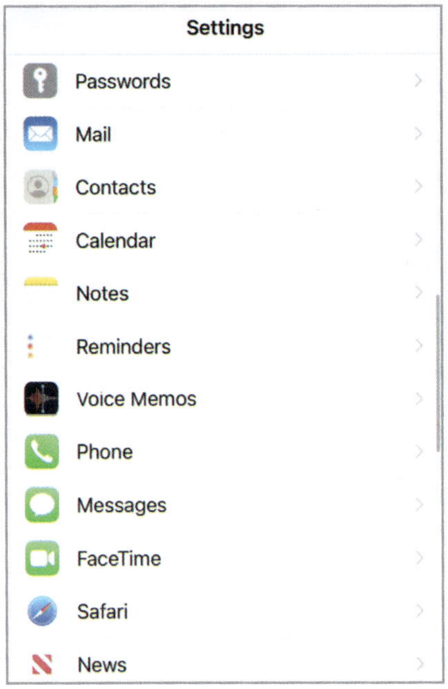

01 Settings

While the Mail app is the front line for your email, a lot of the key decisions are made from within your Settings. So go to the Settings app, scroll down and then select Mail.

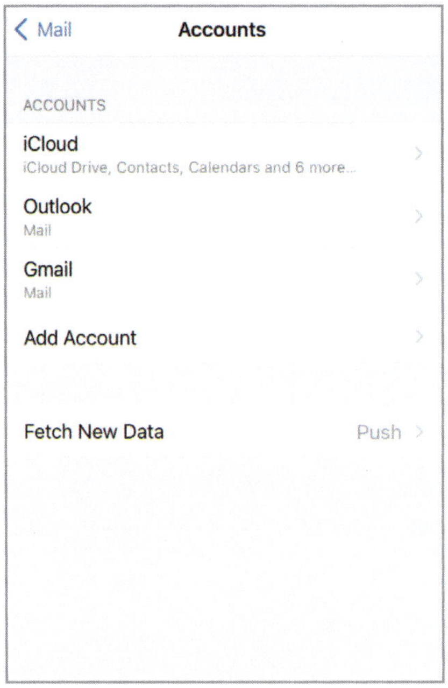

02 Accounts

At the very top of the Accounts section is the list of email accounts you have synced with your device. You can add a new one at any time by tapping Add Account and inputting the details.

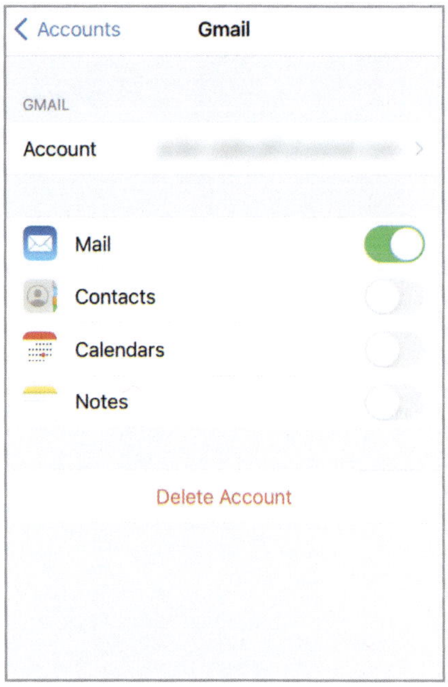

03 Synced features

Tap on any of the accounts in that list and you can toggle the aspects that are synced to your device, so Gmail users can bring across their calendars and contacts, for example.

Inside your default Mail app

Customise your settings to get Mail working as you wish

Accounts
Though there are combined and individual inboxes above, all of your separate accounts can be accessed independently via this part of the app. You'll find all your folders and settings from desktop here too

Compose
Should you need to quickly fire off an email, you can by tapping the Compose icon. If you have multiple accounts signed up for Mail, you can select which account to send the message from

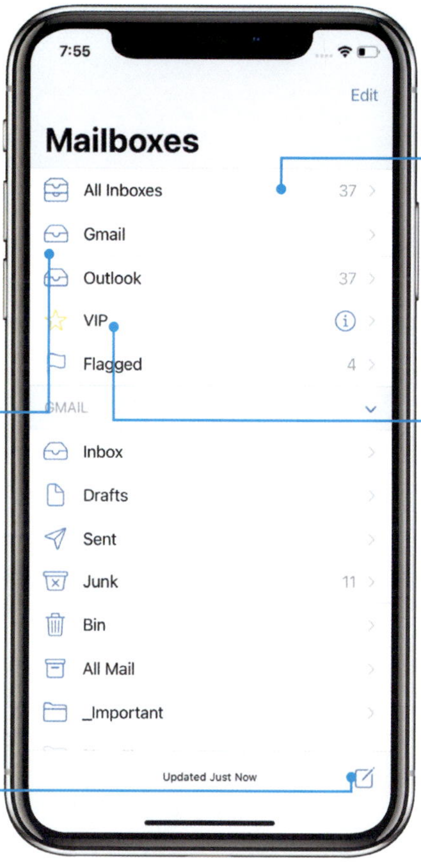

Inboxes
To view just the inbox of an account, tap any of the names next to an inbox icon. Alternatively you can view all your inboxes in one place, as a single list of read and unread emails

VIPs
All mail received from people marked as VIPs will be filtered into their own inbox, making them very hard to miss. Tap on a sender to mark them as a VIP

iCloud
Chances are you will have your iCloud email linked to your Mail app, and with this in place you can also edit your iCloud settings in the same area of the Settings app. This includes adding extra storage should you need it, but also enabling useful features like iCloud Keychain and Find My iPhone to boost security.

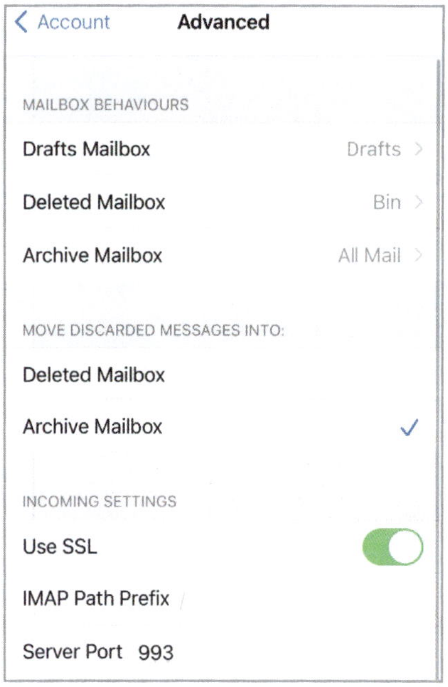

04 Advanced options

Tap on the account address and then Advanced to reach a detailed menu on the behaviour of your inbox. Here you can select which folder an action sends an email to.

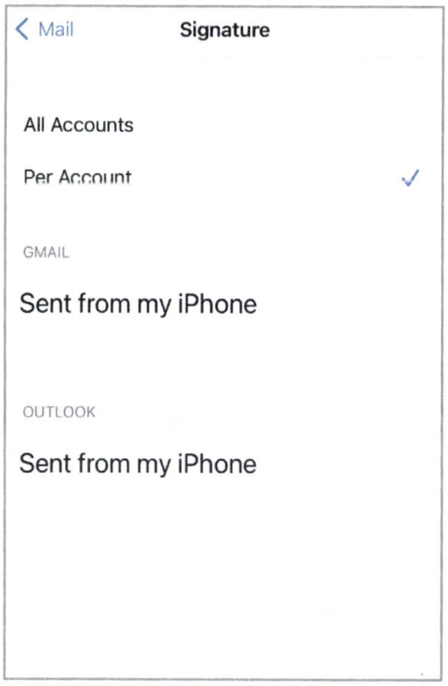

05 Signatures

A lot of us use mobile email for both personal and business reasons, and want our signatures to reflect that. Back on the main Mail page, tap Signature and then Per Account to set sign-offs.

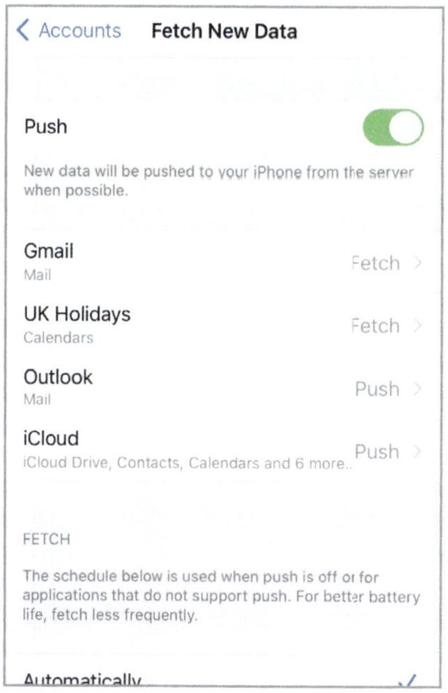

06 Fetch

There are some emails you might be more eager to receive than others. Mail lets you select whether you want each account to push new messages to you, or collect them manually.

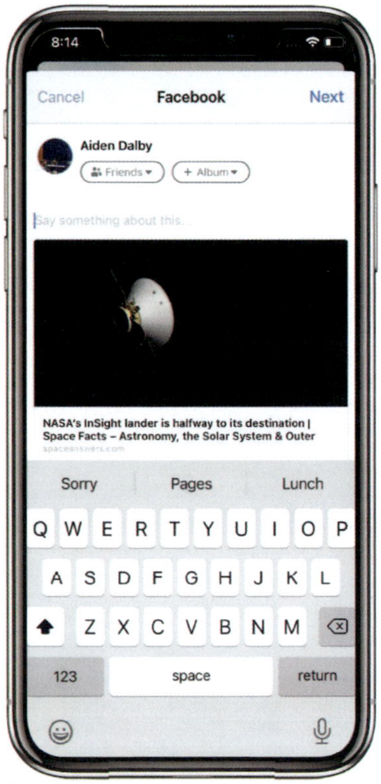

Get more from Safari on your iPhone

It was a tough adjustment moving your browsing experience from desktop to mobile interfaces, but Safari makes it effortless

Browsing the internet on your iPhone is an absolute breeze, and the evolution of iOS over the years has brought even more functionality to the app. With options to block those annoying ads that appear on web pages and save Safari content to your Notes app, browsing the web just got more interesting. Of course, the Safari app also has a whole host of other cool features that make it such a robust browser.

With iCloud Tabs, any tabs that you have open on your iPhone will be viewable on any other device that's running one of the latest operating systems – such as your iPad or Mac. All you have to do is tap the tab button at the bottom-right of the screen in Safari, scroll down past the preview screen images on your iPhone of other open tabs and you will find a list of the webpages that are open on your other devices (at least, those which are connected to the same iCloud account).

The Reading List feature is also useful as Safari saves webpages – not just links – in your Reading List, so you can catch up on your reading even when you can't connect to the internet. It's ideal if you have to endure a long commute to work.

"You can start browsing on one device and continue exactly where you left off on another"

Safari Access tabs, block ads, make notes and more

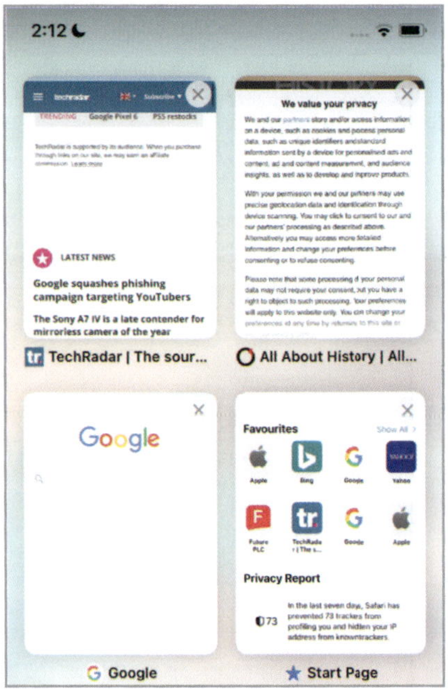

01 Tap on Tabs

While in Safari, tap on the tabs icon, which is to the far-right of the row of icons that sit at the bottom of the interface. All of your open pages will be displayed here.

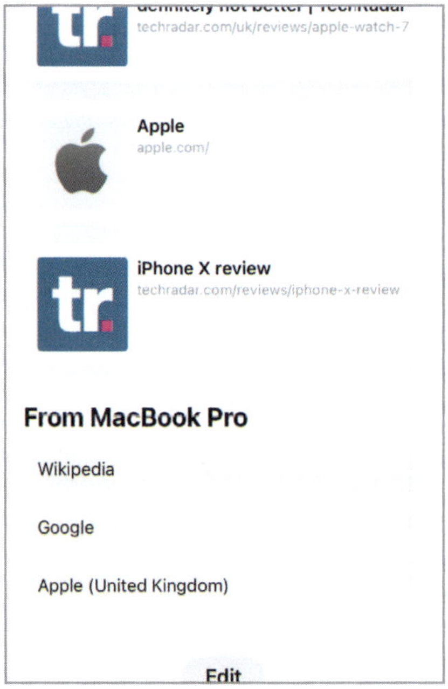

02 Choose iCloud Tabs

When you open a new tab, scroll down to the bottom and you will see all of the pages that you currently have open on your other Apple devices, such as your Mac or iPad.

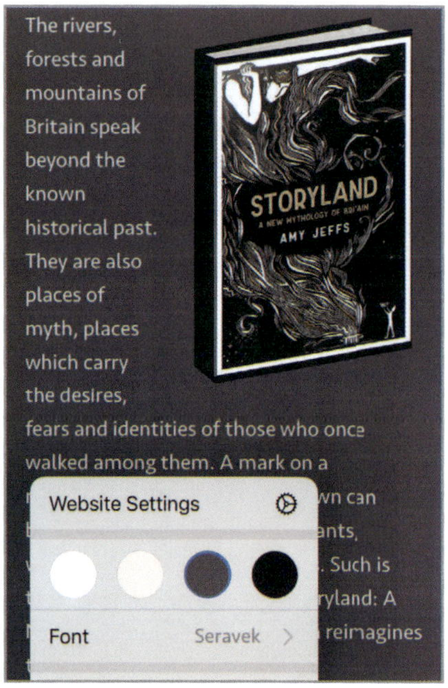

03 Reader View

Tap on the 'AA' icon in the web address bar to enable Reader View. This presents web pages in a different style with customisation options such as font and background colour.

Explore your sharing options

There are new ways to share webpages on your phone

Social network integration

Safari is just one of many apps to feature integration with your favourite social networks. You can post content from Safari just by choosing your preferred service here

Add to Bookmarks

If you access a particular page frequently then choosing this option will save it straight to your Bookmarks. Now whenever you open a new tab you'll be greeted with a handy link right off the bat

Change your search engine

Set to Google by default, you can change the search engine of your Safari web browser by going to the Safari section in the Settings app and tapping on Search Engine. You will then be able to select one of four options, including DuckDuckGo, which is a private search engine.

Notes integration

Create new pages in your Notes app based on Safari content. If you choose Notes from the share menu then the page you are currently on will form the content of a new note, which you can then save to your Notes app. This is very handy when researching

Add to Reading List

By opting to add the current page to your Reading List, Safari will download it, making it accessible via the Bookmarks button to read offline later

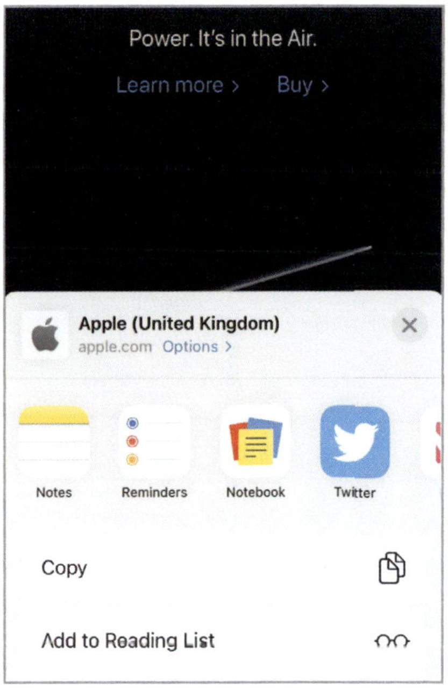

04 Add to Notes

Any content of interest you come across through Safari can be used as the basis for a new page in your Notes app. Tap the share icon, choose the Notes option and a new note will be created.

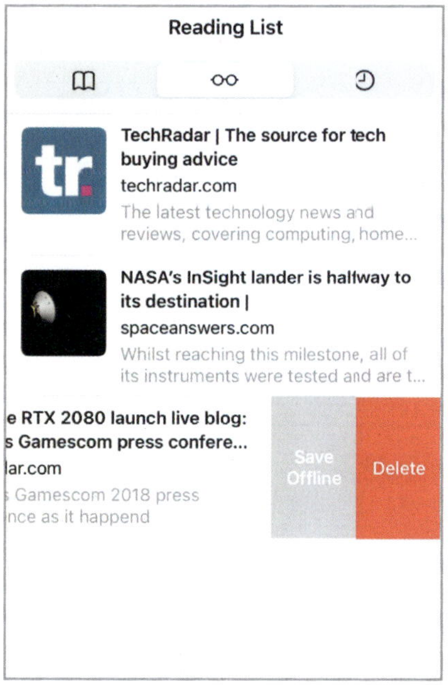

05 Read offline

If you tap the share button and then choose 'Add to Reading List', those pages will then be accessible offline. View them by tapping the Bookmarks icon and then choosing the glasses icon.

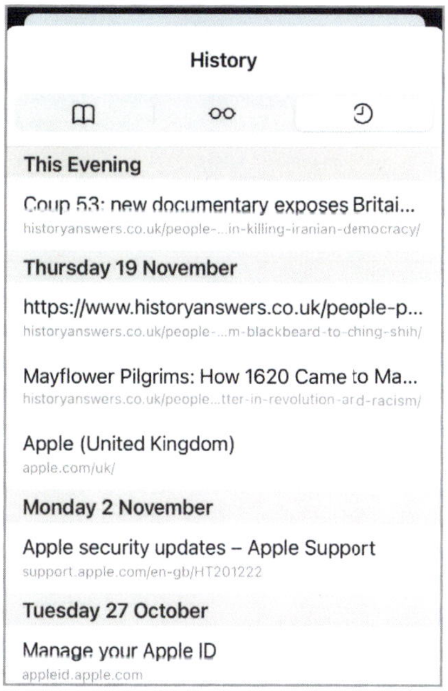

06 Browsing history

You can quickly go back to pages you've visited by tapping on the History tab. This organises web pages by date, so if you remember when you visited a site you can quickly find what you're looking for.

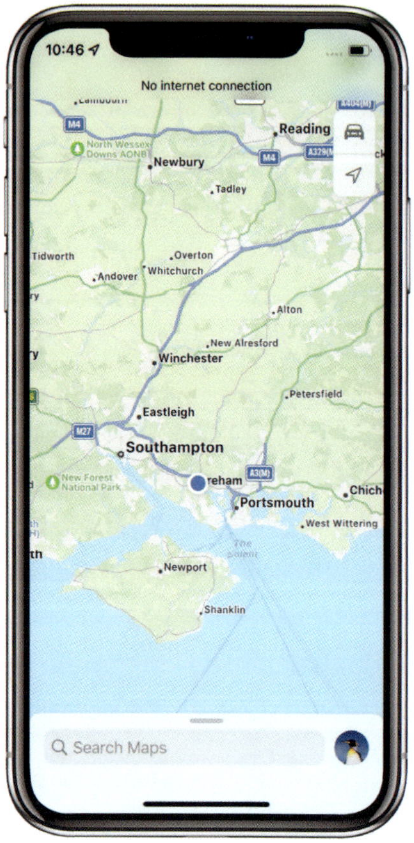

Find your way with the Maps app

Apple's very own Maps app was completely redesigned to keep up with other successful maps applications

Redesigned from the ground up a few years ago, Maps revolutionises the way you view the world. Since the release of iOS 10, all of the navigational items have been moved to the bottom of the screen, which is handy because you can now do all of your navigating using one hand. A drawer at the bottom of the screen slides up to reveal your recent and favourite locations. All you need to do to start finding your way to them from your current location is the tap on the green 'Go' button and the app will immediately generate the directions and present them in turn-by-turn fashion. Also, if you have parked up at a location and have CarPlay in your vehicle then your iPhone will remember and store the exact location of where you parked and help you navigate your way back there later.

Another good advancement of the Maps app is the way in which you can find places like petrol stations and food outlets along your route without having to break away from your journey navigation. The Maps app will find places of interest along the way and allow you to access them by sliding the drawer up.

"You can find petrol stations and food outlets along your route without having to leave the navigation"

Maps Get to grips with Maps

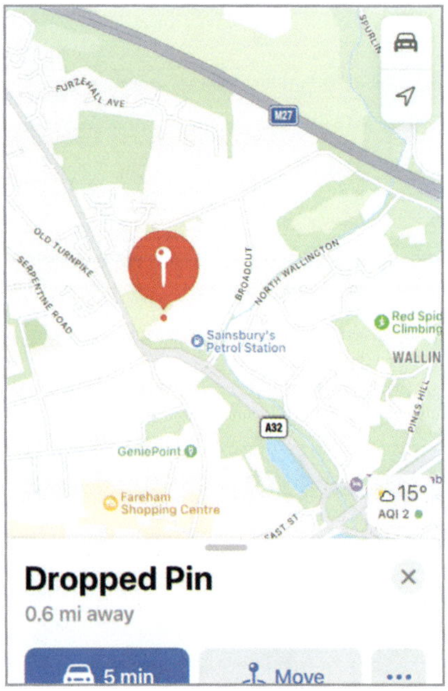

01 Drop a pin

You should use the search field in Maps in order to enter an address or postcode. When the location is displayed, press and hold the screen in order to drop a pin.

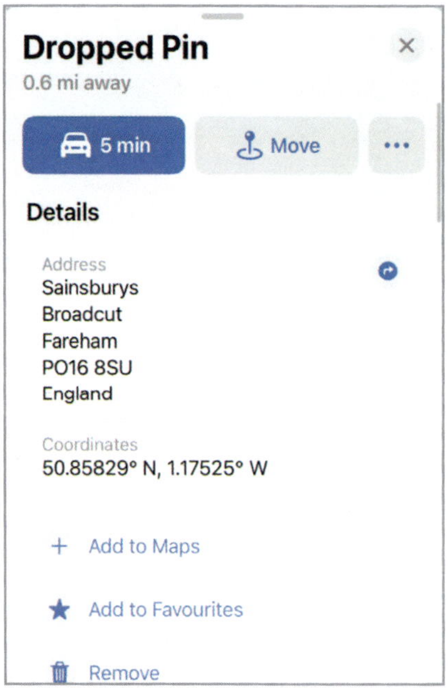

02 Get Info

Tap the pin in order to bring up the name of the location and any available information about the place, plus options to share, add it to Favourites or get directions.

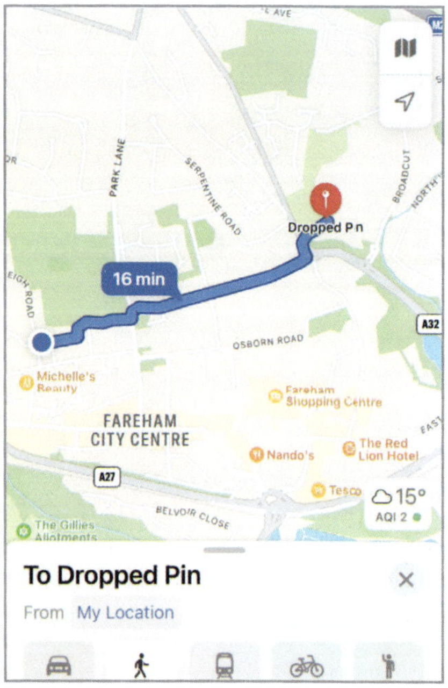

03 Find a route

Tap the Directions button and specify your starting location (it could be your current location or a different one) then the app will calculate various routes for you to take.

Getting there with the Maps app

The Maps app is packed full of options to help you get where you need to go

Turn-by-turn navigation
Spoken directions inform you of every upcoming turn on route to your destination, while 3D view and real-time traffic information will ensure you reach your destination with time to spare

Show Traffic
You can get real-time traffic information to calculate your ETA. Maps gives you details on what's causing a backup, so you can tell if there's a major accident or just a minor delay

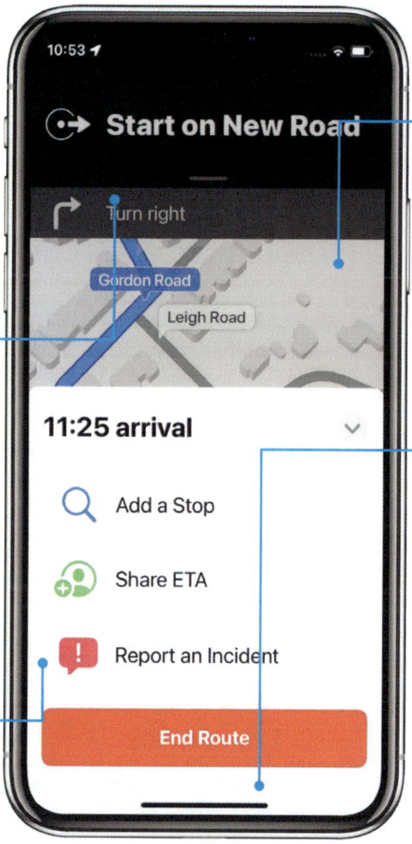

Hands-free directions
Large signs and arrows superimposed over the image show you which way to go and how long it's going to take you to get there, even if your screen is locked

Places of interest
Swipe up to see places of interest along your route. Places like restaurants and petrol stations could be highlighted – which would be particularly useful if you are starting to run a little low on fuel

Flyover
With the Flyover mode, you can see major metro areas from the air with photo-realistic, interactive 3D views. Explore cities in high resolution as you zoom, pan, tilt and rotate around the city and all of its famous landmarks.

04 Start directions

Select your method of transport at the bottom of the screen and then tap on the 'Go' icon next to your preferred route and you will bring up turn-by-turn directions to your specified location.

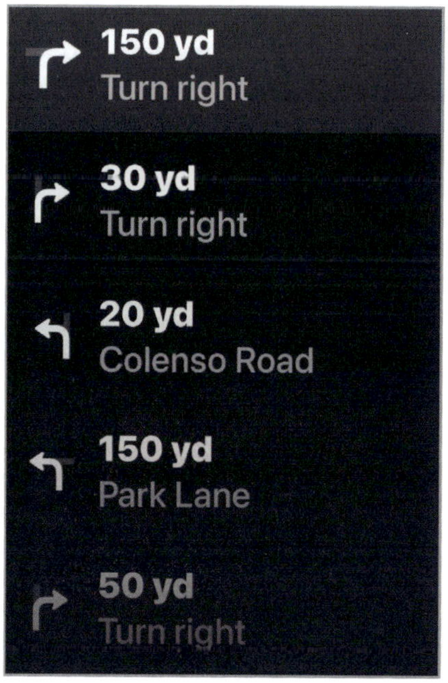

05 See all directions

Swipe up from the bottom of the screen and then tap on the 'Details' option. You will now be able to see of the directions for your chosen route in one scrollable list.

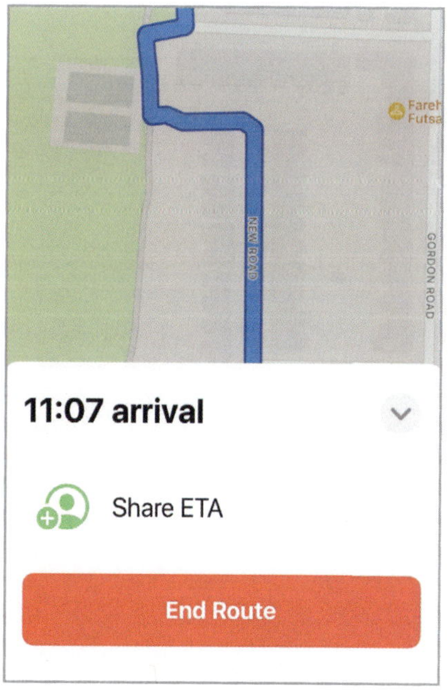

06 Pause your journey

You can also swipe up and use the menu at the bottom of the screen to get an overview of your journey or share your location and ETA with people in your contacts.

Use the Camera

Take better mobile photos than ever before on your iPhone

If there's one category on the iOS App Store that's exploded more than any other over recent years, it's photography apps. There are hundreds, if not thousands of apps dedicated to improving your iPhone's camera and enhancing the stock Camera app but the latest iPhones have plenty of features that make them work well on their own.

The app we know and love today has a simple interface, live photo filters, square and panoramic photo modes and more. There is even a Time-Lapse mode for dynamic, animated photos and a shutter delay function. For creating more dynamic videos, there is the Slow-mo video mode, which can then be edited in the Photos app for tweaking.

If you love adding filters and effects, and the medium format-style images that Instagram gives you, you'll love what Apple has done with your Camera app. There are tons of filters to choose from, not including the standard, filter-less image, but if you want a quick result our favourites are Noir (for black and white shots) and Dramatic Warm (for a Polaroid-style look).

Here we provide a quick tour of the Camera app's key features. No matter whether you're an absolute beginner or well-versed in iPhoneography, we're sure there is still a trick or two to learn to ensure you make the most of your treasured snaps. All our favourite features are detailed below, so get started now for flawless photos.

Camera Rediscover your iPhone's camera

01 Open it up

To fire up Camera, select its icon from the home screen, access it from Control Centre or swipe up from the bottom-right of your lock screen. By default, the app will load in standard Photo mode.

02 Slide to switch

Switching between modes is easy; swipe left or right on the mode name between the shutter button and viewfinder. Square mode gives medium format-style images.

03 Panorama mode

Swiping the to the left once more will put you in Panorama mode – simply press the shutter button and follow the on-screen instructions to create a super-wide image.

Advanced camera tricks

Extra advice for perfect shots

Live Photos
When Live Photos is turned on, the camera will capture a short video at the same time it captures the picture. Go to the Photos app to edit these, there are three effects to choose from

Check your images
To view your images in Photos, tap on the image in the bottom-left corner of the Camera interface. It's a handy shortcut to your most recent snaps and videos

Shutter delay
Want to line up a shot and get in on the action yourself? Then tap here and set a shutter delay of three or ten seconds to give you enough time to get in the frame

Hold to focus
You can tap anywhere within the Camera viewfinder to set your focus, but you can also lock both focus and exposure by tapping and holding for a few seconds

Going for HDR?
If you're shooting HDR images and want to save space on your iPhone, be sure that the Keep Normal Photo setting is turned off under Camera in the Settings app. This will stop the Camera app from saving an HDR and a regular version of your image every time you take a snap.

04 Add some effects

To access the Camera app's live filters, tap the icon in the bottom-right corner (this is greyscale when no filters are applied and in colour when they are) then select one.

05 Take a Time-Lapse video

Time-Lapse mode is a nifty feature that will capture shots at dynamic intervals to create a still video with slight movement. Try experimenting with this for great results.

06 Video mode

From the standard Photo mode, swipe right once to turn your camera into a video recorder. You can't add live filters in this mode just yet, nor shoot your clips in HDR.

Understand the powerful Camera modes

The iPhone cameras are not only technically brilliant; they also provide a varied selection of powerful modes to play with

The iPhone 12 and 12 Pro cameras are extremely impressive and you should have all of the tools you need for any type of photo or video built into the software. We will show you how to use each of the included modes here and how to add some personality to your shots, but as with any camera, patience and a basic understanding of lighting will always help achieve the best photographs. It's highly likely that your iPhone is good enough for you to leave your standalone camera at home, and for most purposes it will produce shots that are more than clear enough in every type of situation. If you have an iPhone X or newer then your camera has the Portrait Lighting feature, which creates studio lighting effects to provide the best selfies.

"The iPhone is good enough for you to leave your standalone camera at home"

Extra functions
At the top you can choose whether to enable HDR or not, set a timer for group shots and also turn the flash on and off

Be focused
Tap the screen to focus and slide your finger up and down to adjust the exposure. Every adjustment is previewed in real-time

Camera Take control of the iPhone camera

01 Lighting and timing

Before you snap, check the options at the top. You can set the flash to automatic, on or off. To enable High Dynamic Range (HDR) photos you must go into the Camera section in the Settings app.

02 Tap to focus

With 'PHOTO' selected at the bottom of the screen, tap anywhere and the lens will focus on that area. This means you can make any part of the foreground the subject.

03 Control the exposure

When focussed, you will see a sun icon next to the yellow square. You can move this up and down to manually control the exposure and your changes will be shown in real-time before the photo is taken.

04 A perfect square

Scroll the modes to 'SQUARE' at the bottom to use the camera in the same way as normal, but the resultant photo will be formatted for social networks like Instagram. It's a surprisingly useful mode.

05 Capture a panoramic photo

The 'PANO' option is very clever, but you will need a steady hand. When in this mode, tap the capture button and slowly move the iPhone from left to right; the end result will be impressive.

06 Capture a video

Capture a video by scrolling to 'VIDEO' mode and hitting the capture button. Try to take videos while holding the iPhone horizontally as the end result will be more suitable for YouTube.

07 Slow it down

Once you have selected the 'SLO-MO' mode and taken a video, you can play it back and slide the top bar to choose exactly where the slow motion effect starts and ends.

08 Time lapsing

Select 'TIME-LAPSE' and then the capture button to take a series of shots which will display as a time-lapse once complete. You need to take a long video because the final product will be much shorter.

09 Clever filters

Tap the icon in the camera mode that looks like three interlocked circles. You can now access filters which will preview in real-time and be applied to your photo. It's easy to remove the filter afterwards.

Shoot video on your iPhone

You probably didn't buy your iPhone to record video, but it's a valuable function that shouldn't be overlooked

The old photography truism that the best camera is the one you have with you applies as much to shooting video as it does to taking snapshots. You might be browsing the web, reading a book or listening to music on your iPhone, but within seconds you can use it to quickly grab priceless footage of that never-to-be-repeated moment.

In fact, the iPhone is equipped with two cameras: one front and one rear-facing. And while you'll probably get a lot of use out of the front-facing camera thanks to the FaceTime app, the better quality of the rear camera makes it more suitable for recording video with.

Accessing and using your iPhone's camera is easier than ever as you can access it from the Control Centre or directly from your iPhone's Lock screen. To do this, simply go to the Lock screen and a camera icon will appear next to the slider that you use to unlock your phone – tap this and your phone will unlock and launch your Camera app simultaneously to ensure you can start recording straight away – a real bonus when you want to capture a spur-of-the-moment video.

What's more, once you have captured video, you can edit it straight away from your device using the Photos app, which allows you to trim your footage down to contain just the important stuff. After that, you can share footage with friends within seconds via Facebook or email. In this tutorial we guide you through the process of capturing, editing and sharing videos.

Camera Create a video with your iPhone

01 Launch the video

Tap the Camera app's icon on your iPhone's Home screen to launch it. Just above the shutter button will be various different modes available. Swipe left to choose Video.

02 Recording video

Tap the record button to begin recording. As you record, a red dot will slowly pulse at the top of the screen, and at the top of the screen will be the elapsed time of your video.

03 Change focus or exposure

To change the focus of your video or set its exposure, tap at the point where you want to set the focus or the exposure, and the rest of the video will adjust to compensate.

Record cool video footage

How to control your camera

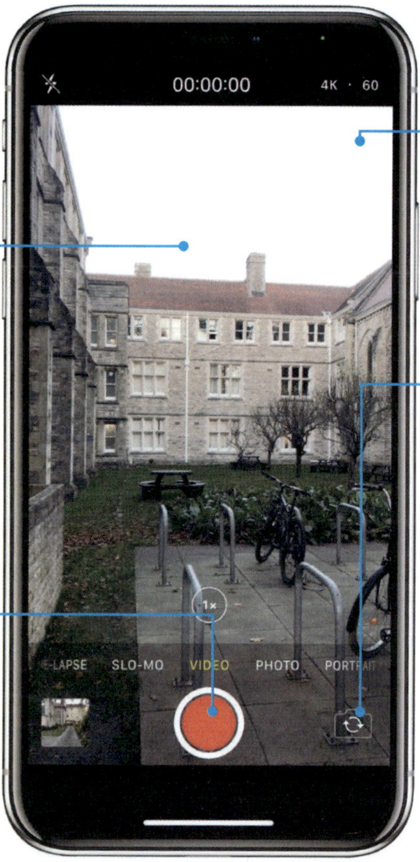

Lock focus

You can lock the focus and exposure by simply tapping and holding in the viewfinder until the focus box pulses and shows AE Lock

Camera quality

You'll get better results using the rear-facing camera, which can record 4K video at up to 60 frames per second.

Front or back

When you're not recording, an icon appears in the lower-right corner of the screen that lets you switch between the front and rear cameras

Control recordings

The record button pulses as you record and the viewfinder displays the elapsed recording time. Just tap the record button to stop recording

Take snaps on the fly

While recording video footage on your iPhone, you will notice a second button to the left of the main record/stop record button. This is the camera shutter and if you tap it while you are recording then you will instantly take a picture at the same time. This is a great way to ensure moments are well and truly captured!

04 Stop recording

Tap the record button again to turn recording off. A thumbnail link to your freshly created video appears at the bottom-left of the screen. Tap it to view the video you've just taken.

05 Edit the video

You can trim the video by tapping on the edit button in the Photos app, dragging the start and end points on the filmstrip at the bottom of the screen. When you're happy, tap the Done button.

06 Share with others

To share your completed video with others, tap the share icon in the bottom-left corner of the screen, select your video, tap Next and then choose a means of sharing.

Explore the Photos app

Quickly organise and arrange your images thanks to your iPhone's GPS data and Photos app

With the iPhone's camera just as good as any compact point-and-shoot digital camera and far more convenient to fire up at a moment's notice, it's no surprise that our iPhone albums get quickly filled with photos.

On the most basic level, your photos are sorted year-by-year with the date and time of an image determining which collection it's sorted in. Things can get even more organised too, with photos being arranged in categories such as faces and places, so that you can always find a particular image quickly.

Once you've found that perfect photo, you can edit it using Photos' built-in tools or share it via a number of different methods including AirDrop. If you wish to make quick edits to your photos on the fly, then tap on the photo thumbnail in the lower-left corner of the Camera app and you will be able to get started editing and sharing straight away – it's really easy.

Another great feature of the Photos app is the For You section. As the app utilises the artificial intelligence of Siri a lot more, photos, live photos and videos from a certain event are now brought together into an interactive magazine format, so if you're feeling nostalgic, you can sit down and experience all of the joy that a particular event brought you again and again.

Photos View your snaps in style

01 Year by year
On the most basic level, Photos organises your iPhone photos by the year they were taken. If you're a real shutterbug, this should help you narrow things down.

02 Understand Collections
For You works by grouping photos by the date they were taken and the general location they were taken in – they're perfect for quickly sorting through your photos.

03 Work with Albums
Albums detects common themes in your photos and groups them into collections such as location, date, people and more. You can also find your recently deleted images here.

Treasure your memories

How to relive those past precious moments

Recorded and dated
Each For You album is date stamped so that you will know, at-a-glance, exactly when it took place. The more images and videos you capture, the more albums will be recorded

Intelligent groups
Using Siri's AI brain, the Photos app plucks your best images from an event, location, etc, and groups them all together into one pace

Your Memories
When you tap to view a For You collecyion you will see an automatically generated video at the top. This is like a highlights reel that showcases all of the best photos, live photos and videos from the accompanying album. You can go into the options to change the theme and adjust the length of the video to suit your needs.

An interactive memory magazine
Memories brings together videos and pictures and presents them like an interactive magazine that you can delve into time and again to relive those precious memories

New section
Your personal Memories will be accessible via a new section tab at the bottom of the interface. Go here to see your treasured memories turned into interactive albums brimming with charm

04 On the map

Tapping on the Places option will reveal a map view showing images you've taken sorted by their location on the map. They will show up with the number of images taken there.

05 Single images

Once you've found the image you want, tap it once to show it full screen and tap it again to lose the Photos interface for a clutter-free viewing. When done, tap to bring the menu up.

06 Edit your images

Tapping the Edit icon will bring up a number of familiar editing tools along with a collection of photo filters to explore. You can really start having fun with your photos here.

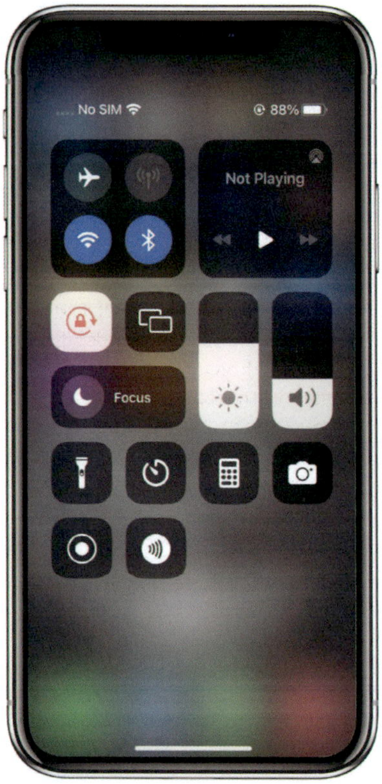

Multitask with the Control Centre

Switch between apps with ease thanks to the multitasking feature and Control Centre

When it was first introduced, Control Centre was a popular addition to the operating system as it allowed you to access some of your iPhone's most common settings and features quickly. In iOS 11 the Control Centre app received a massive redesign. The media controls, which when you had music or a movie playing were previously in a separate tab, now sit in the same panel as the rest of Control Centre's commands. Also, if you have been using the Home app to link any compatible appliances in your home, such as cookers, lights and home security systems, to your iPhone, Control Centre will allow you to access and control them instantly. Pressing and holding on certain icons will bring up more controls such as the torch. Holding on this icon will bring up a brightness slider.

The multitasking aspect of your phone, in which you can effortlessly switch between open apps on your device and, if necessary, close them down, remains the same as ever. Simply swipe up from the bottom of the screen to reveal your open apps and then scroll left or right through them and swipe up to close them. Read on to find out more about Control Centre and multitasking on your iPhone and how to use these features.

> "Effortlessly switch between open apps on your device and, if necessary, close them down"

Control Centre Learn to use the panel

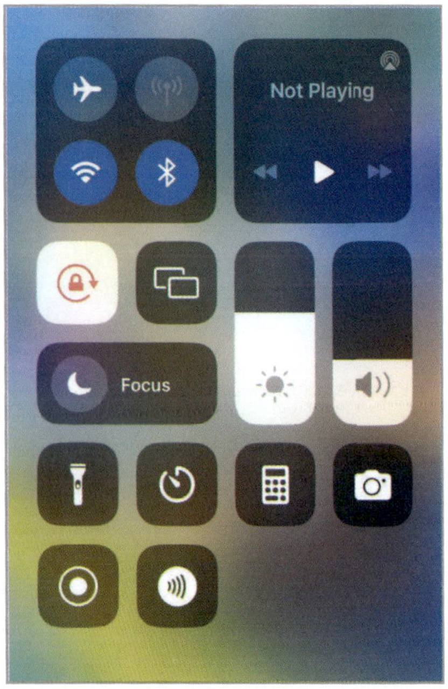

01 Swipe to start

To access Control Centre, simply swipe down with one finger from the top-right corner of your iPhone's display. The panel that appears will reveal the quick access controls.

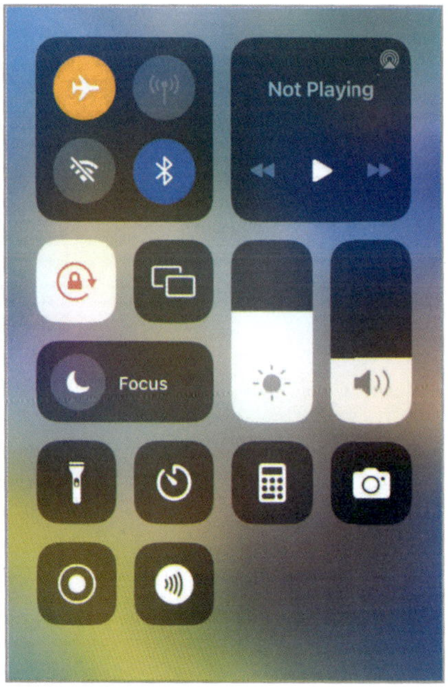

02 Tap and enable

Some of the most commonly used settings such as Airplane Mode, Bluetooth and Wi-Fi are at the top of Control Centre. Tap their icons once to enable or tap and hold to see more.

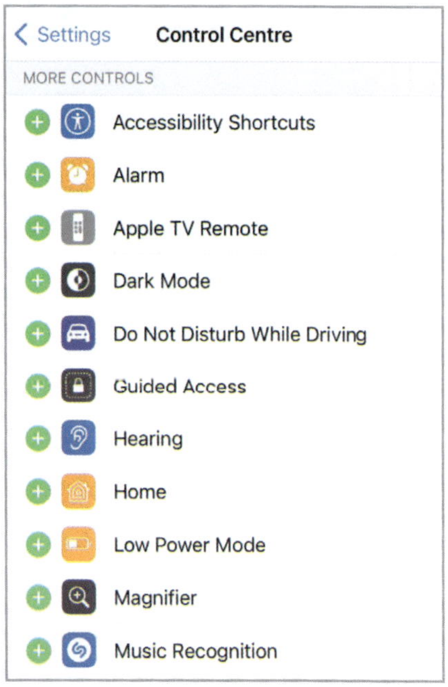

03 Expand Control Centre

To add more app shortcuts, go to Settings> Control Centre and you will see a list of controls. Tap on the green '+' icon to add a control or the red '-' icon to remove one.

Get more from Control Centre

Explore a few more of the available options

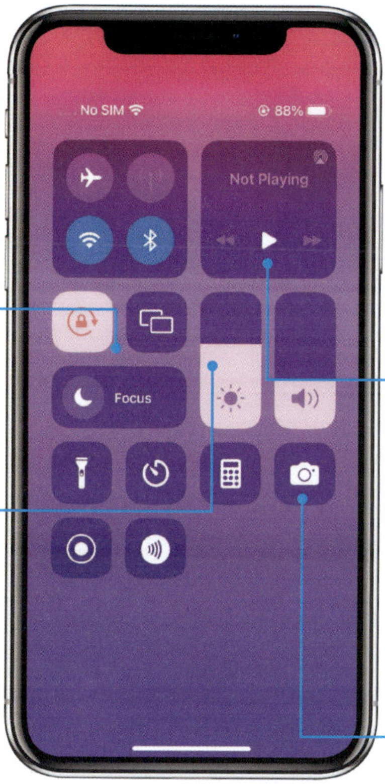

From any app
Control Center is a universal feature from iOS 7 onwards, so it can be accessed from any app. You can stop this in Settings, though, to avoid interrupting activities such as playing games

Settings sliders
When your iPhone's automatic brightness controls appear to be failing you, use the slider in Control Centre to ensure that you're always able to see what's on the screen

A multitasking trick
If you're desperate to shut down a number of apps quickly in the multitasking view, use two or three fingers to swipe up on two or three different app previews at the same time – this secret multi-touch gesture will save you minutes every week of the year.

Play the music!
If you've got music playing, whether it's in Apple's Music app or a third-party offering, the controls appear on the Control Centre page. Tap and hold this for more options.

Quick access apps
The row of buttons along the bottom of Control Centre will take you straight to those key apps such as Camera and calculator. More can be added by going into the Settings app

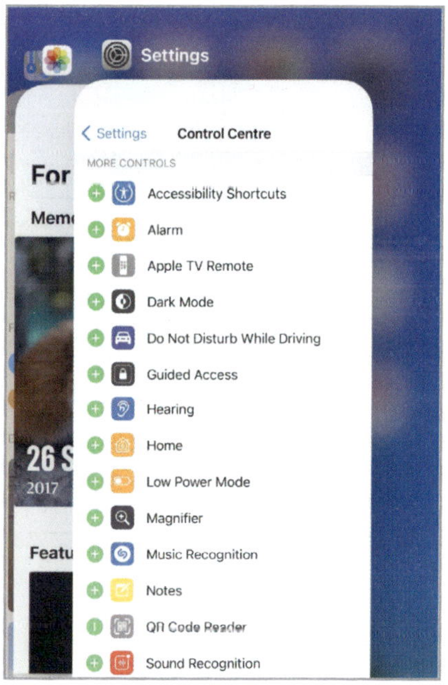

04 Tap it twice

To access the multitasking view, swipe up from the bottom of the screen. If you are using an older iPhone or an iPhone SE you can access this by double-tapping the Home button.

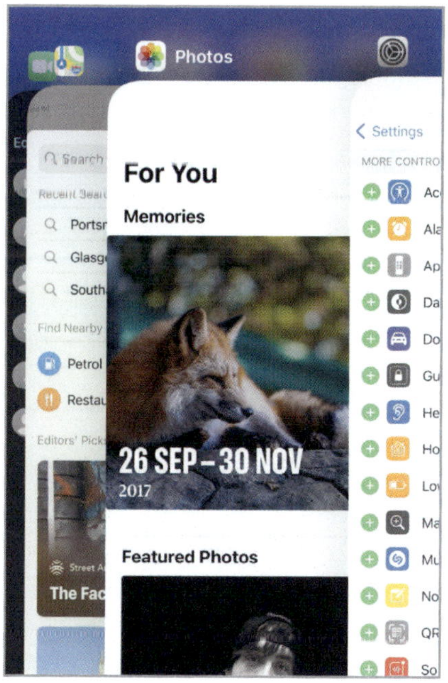

05 Swipe to switch

To move between apps, swipe from left to right in the multitasking view. You can see a preview of each app that's open and tapping on it will take you into the app.

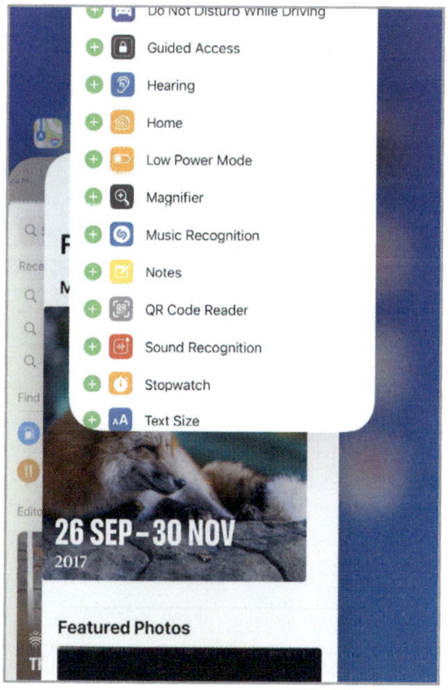

06 Quit the app

To shut down the app, swipe the preview up towards the top of the screen with one finger and 'flick' it away. Do this regularly with unused apps for better battery life.

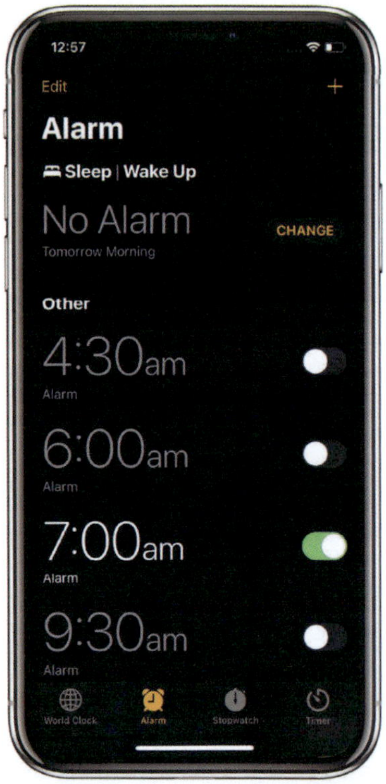

Tell the time with the Clock app

Keep track of the time anywhere in the world and never be late again with the feature-packed Clock app

Your iPhone's Clock app is incredibly versatile. When you launch it you will see a row of icons along the bottom of the screen that represent the various features of the app. World Clock lets you see what time it is in any location across the world, Stopwatch lets you time events and Timer lets you set a countdown timer, which is especially handy for cooking. The Alarm function is what we will concentrate on in this tutorial, helping you set an alarm for a wide range of purposes, but it is also worth pointing out the brand-new feature of the Clock app that was introduced with iOS 10 – Bedtime. As we all benefit from getting a consistent amount of sleep every night, the purpose of this feature is to work out how much sleep you need, determine when you need to get up and then, based on that information, set a reminder of when you need to go to bed in order to achieve it. While it may sound as though your iPhone is turning into a nagging parent, sometimes we all need reminding that we can watch that extra episode our favourite show tomorrow and that in order to function properly, we should really get a good night's sleep!

Clock How to set alarms

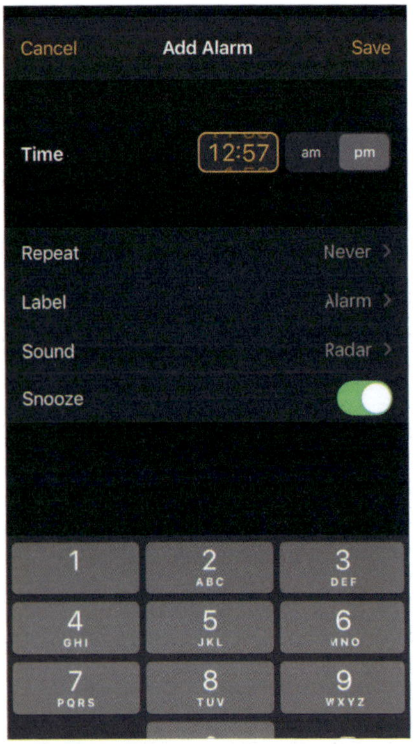

01 Set an alarm

To set an alarm, tap the '+' icon in the top right-hand corner and then use the number pad to configure the exact time or scroll up and down using your fingers until you find the right one.

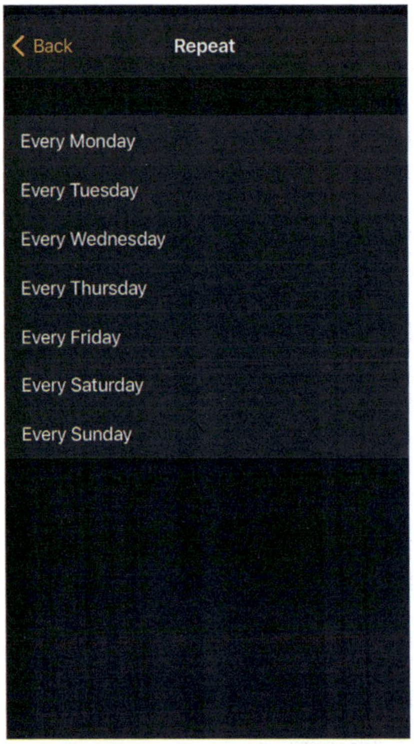

02 Repeat and snooze

You can then decide whether or not you want to repeat the alarm, which you can do on any day of the week, and whether or not you would like to include a snooze function.

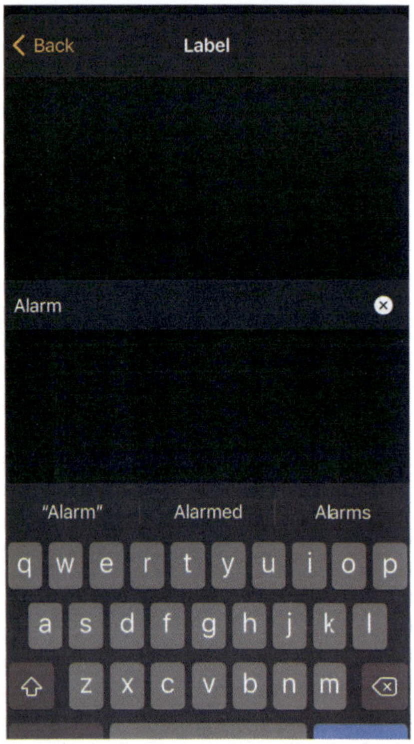

03 Add custom names

It is possible to label that particular alarm with any name you wish. For example, you could decide that you would like to label one 'Waking up' and another one 'Meeting'.

Rest-up with the Bedtime feature

Setting up the feature designed to help you get the optimum amount of shut-eye

Sleep alarm
Based on the amount of sleep that you think you need, the app will count back from the time you need to wake up and then sound an alarm when it is time to go to bed. The idea being that you get a consistent amount of sleep

Your sleep laid bare
When you have set up the Clock app's Bedtime feature, you will be able to clearly see your optimum amount of sleep presented as a meter on the screen. Presumably, if you were awake during the time you should be asleep then you will see this meter deplete throughout the night

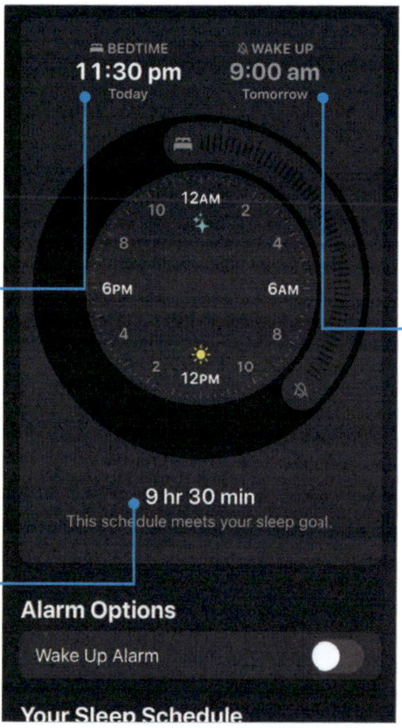

Sleep Schedule
At the bottom of the screen will be a link to edit your sleep Schedule. This will take you to the Health app where you can enter details such as setting your Sleep Goal and Wind Down times. This data will sync with the Clock app.

Wake alarm
When setting up the Bedtime feature, you will need to first determine the time you need to get up in the morning. Once done, the app will count back based on the amount of sleep you have stated that you need

Stop the music
A clever feature of the Clock's Timer function allows you to turn off any music you may be playing as you drift off to sleep automatically after a certain amount of time. Tap on the 'When Timer Ends' option, scroll down to the list of options and then select the 'Stop Playing' option. Now set the timer for however long you want.

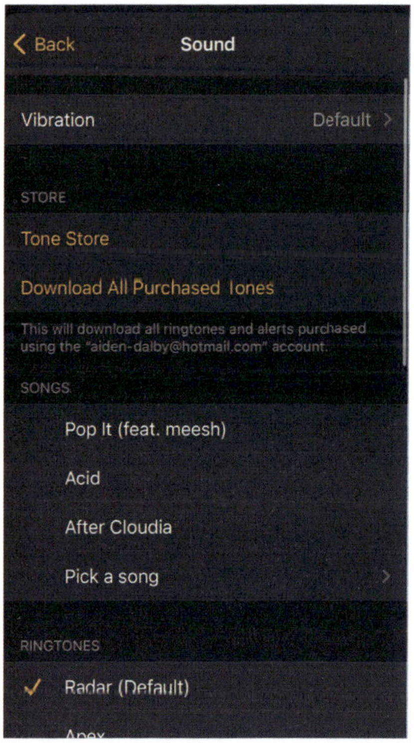

04 Tailor the sound

Tap on the Sound option to pick a pre-loaded alert noise or scroll to the top to pick a song from your Music library. You will also see the option to buy Tones, which takes you to the iTunes Store.

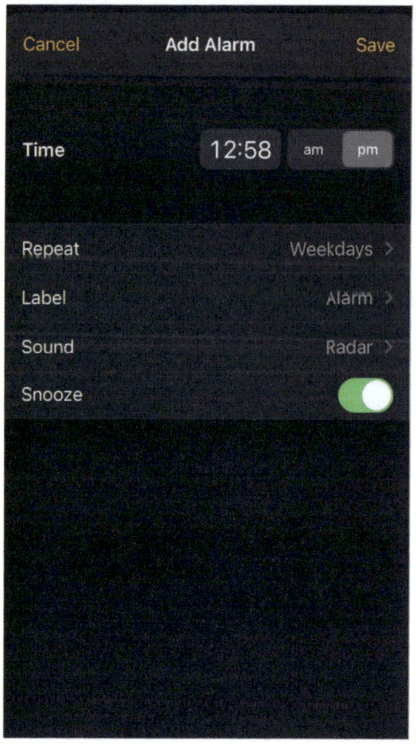

05 Save your alarm

When you have tweaked everything accordingly, tap the Save option in the top-right corner of the screen and your new alarm will be added to the alarms list where you can turn it on and off.

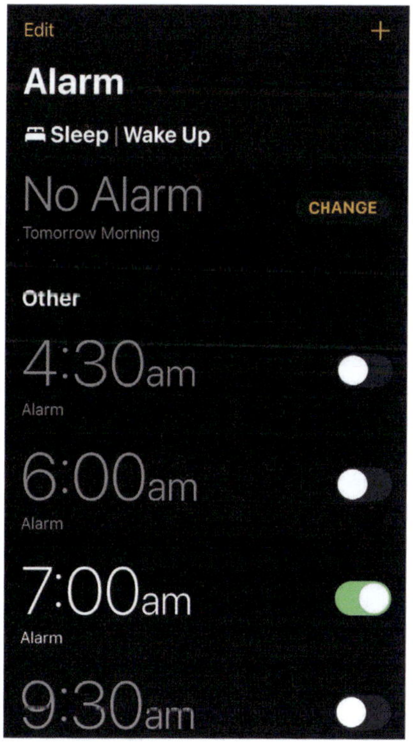

06 Edit alarms

If you wish to disable an alarm then you can swipe the green switch next to the alarm to the 'Off' position. To remove it from the list entirely, tap Edit and then hit the red symbol next to the alarm.

Add an event in your Calendar

Make sure you never miss an important event with Calendar on your iPhone

Tuesday 10 **The Calendar app on the iPhone is pretty useful and very easy to use.** Of course, this digital calendar has a multitude of advantages over a real one. Firstly, you get the beauty of typeface rather than scrawled handwriting; secondly, there's no need for Tipp-Ex; and thirdly, you can list your appointments in a number of different ways – List, Day or Month views are available – to make everything easy to take in.

Like the majority of iPhone apps, the Calendar app is easy to use; so easy, in fact, that you'll want to document every move you make while using it, from eating breakfast to scheduling tea breaks. Adding an event is simplicity itself, and your events will sync with iCloud to be made accessible on all of your devices.

"Calendar is so easy to use that you'll want to document every move you make while using it"

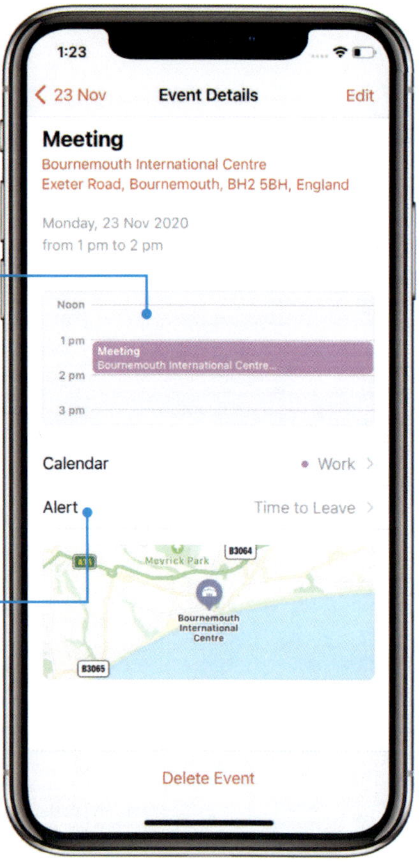

Duration
With the All-day slider set to Off you can set start and end times for events. Flick it to On for a day-

Alert
Notifications are synced with iCal or Outlook on your desktop, so you'll be reminded there as well

Calendar Add an event to your digital calendar

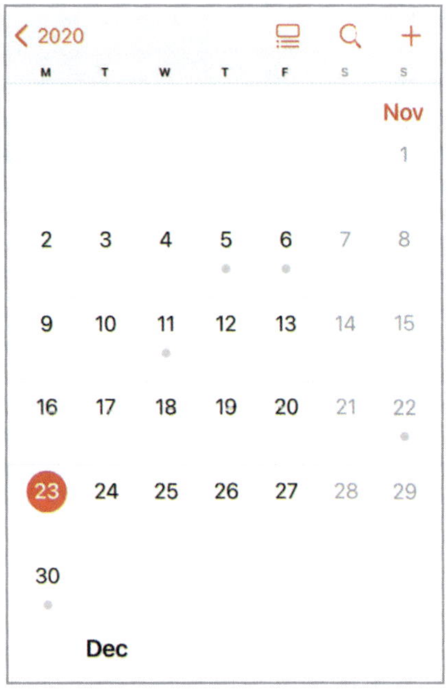

01 Open the app

Open the Calendar app on your phone and select the date that you want. Presumably, you will want to assign an event, so tap the '+' button on the top right of the screen.

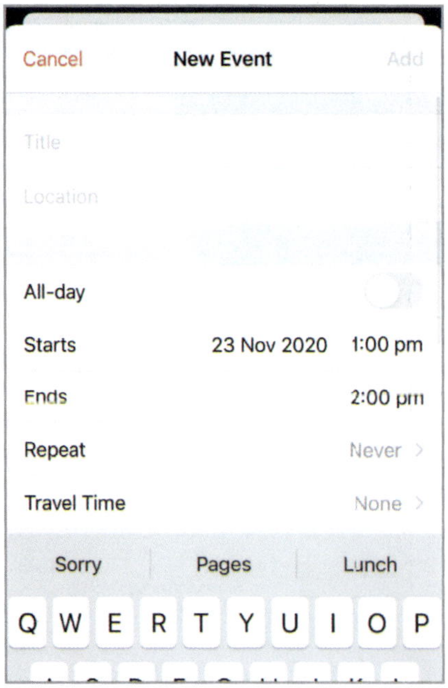

02 Add event

After selecting a date, the above menu screen will appear on your phone, with various time slots, in which you are able to edit all of the details of the new event.

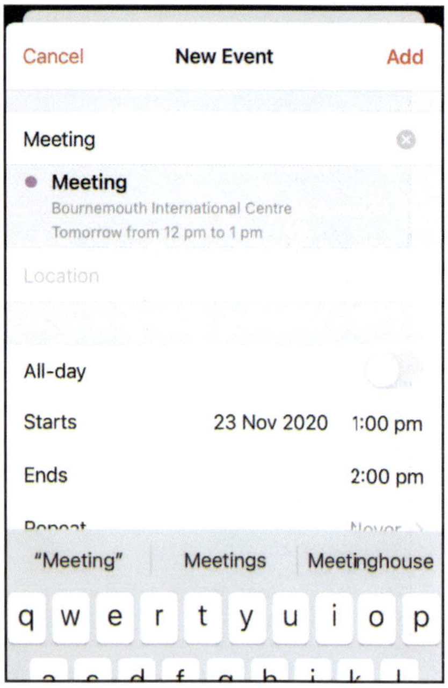

03 Add details

You can add as much or as little detail as you would like to about the event. Such as time, repetition, travel time, locations, people who will be attendance and whether to mark you as busy.

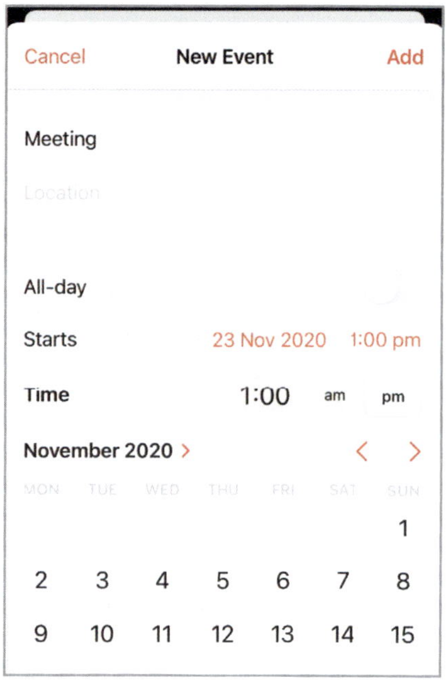

04 Time it

After adding all of the details, you now need to add the start and end dates of your event to the calendar. Tap on the relevant field in order to see the menu.

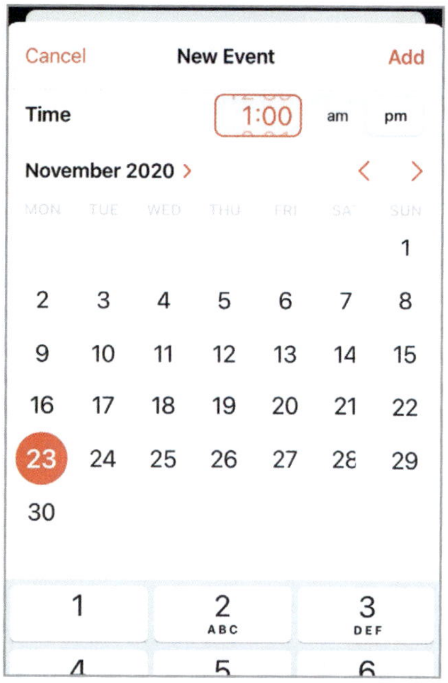

05 Use the number pad

Make sure you use the number pad in order to select the exact times and dates that you want to use. Alternatively, you have the option of toggling the All-day option to 'On'.

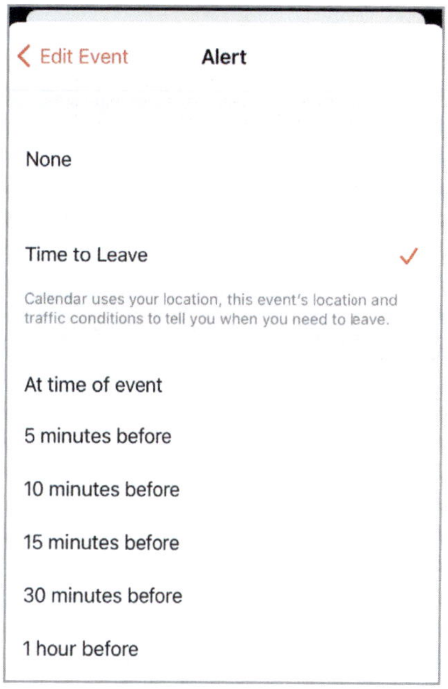

06 Set an alert

If you would like to set an alert for the event, tap the Alert field to set a reminder. These will pop up on your iPhone at the times that you set them, so you won't forget your event.

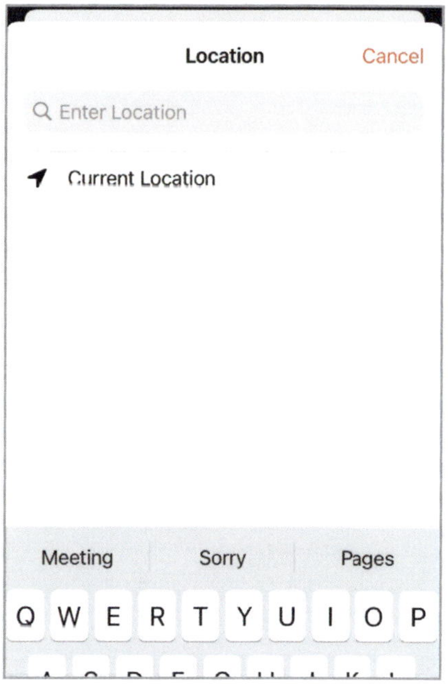

07 Add a location

If your meeting, appointment or errand takes you to a specific place, you can add a location. Beneath the name of the event, tap the faded out 'Location' and search. Tap on the location to add it.

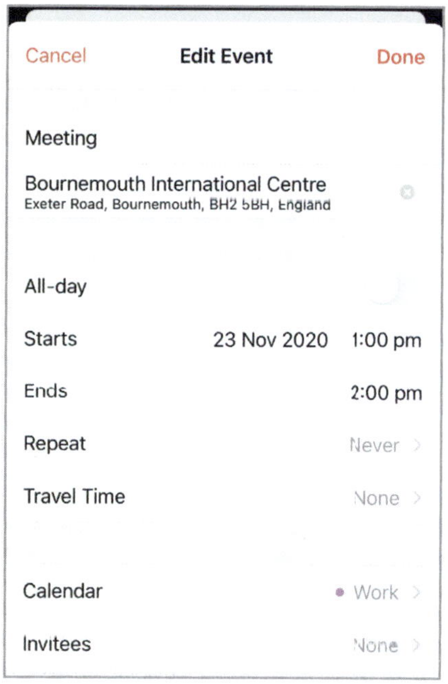

08 Tap save it

Tap on the option that you would like to use. You are able to save your progress by tapping the Add button that appears on the top right-hand corner of the screen.

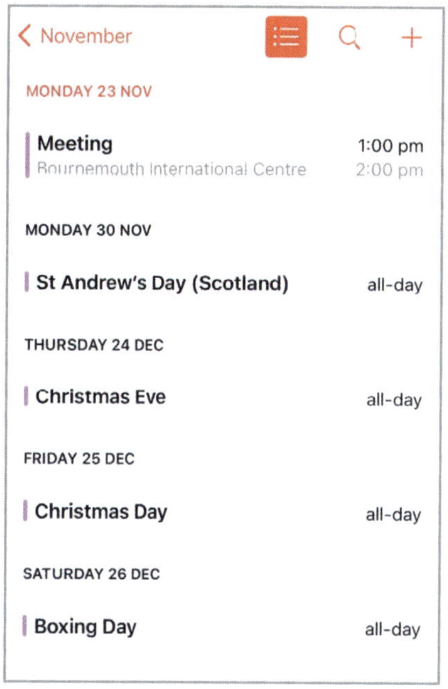

09 Save and view

Your event will now appear as its own separate entry on the page. Tap on it to see the full details and tap Edit to make any further changes or delete the event.

Set up location-based reminders

You can create reminders based on specific locations with ease thanks to the Reminders app

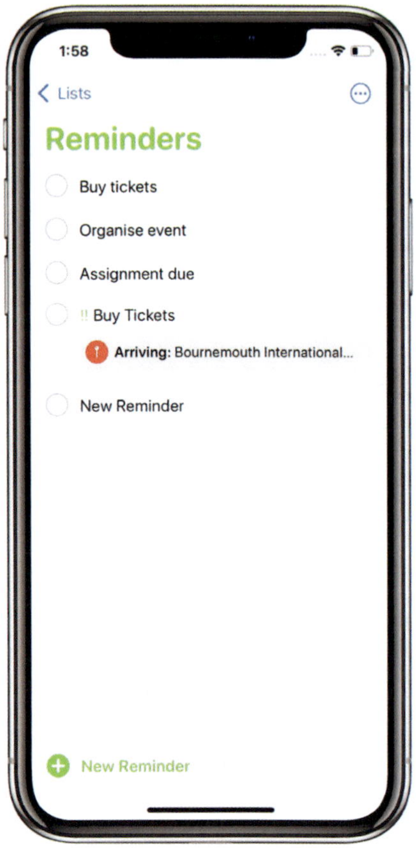

The iPhone Reminders app is one of the simplest solutions offered in all of iOS. Apple has taken great care to make the interface as sparse as possible and the flexibility included has been kept deliberately limited so that you can set your reminders and then forget about them until the time comes when you need to be alerted again.

However, there is a hidden feature that is very powerful and which you will likely use many times once you realise just how beneficial it can be. The ability to set reminders based on a specific location opens doors to automating many of the tasks you need to complete.

There is no better way to complete a task than having it pop up right in front of you when you are in a position to deal with it and this is why location-based reminders are so powerful. They can simplify much of your daily organisational needs, as we'll show in this tutorial.

This app will prevent those moments when you arrive home only to remember you were meant to buy bread – made even more annoying by the fact you were at the shop!

> "There is no better way to complete a task than having it pop up right in front of you when you are in a position to deal with it"

Reminders Set location-based reminders

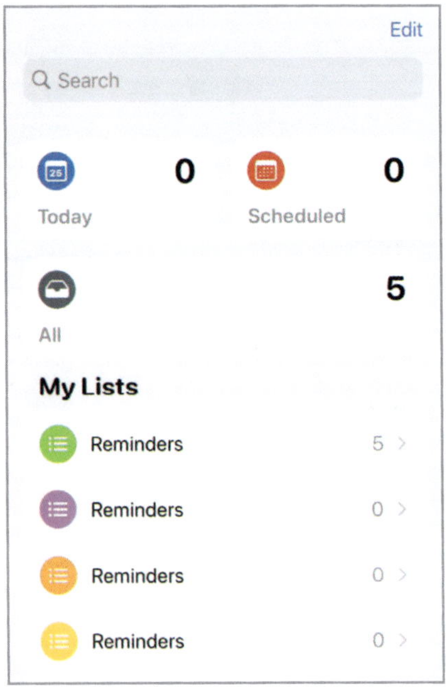

01 Create a reminder

From the main screen tap on the Reminders tab, which brings up a list of reminders. You can also tap on Add List to create a new set of reminders. Open a list and tap New Reminder.

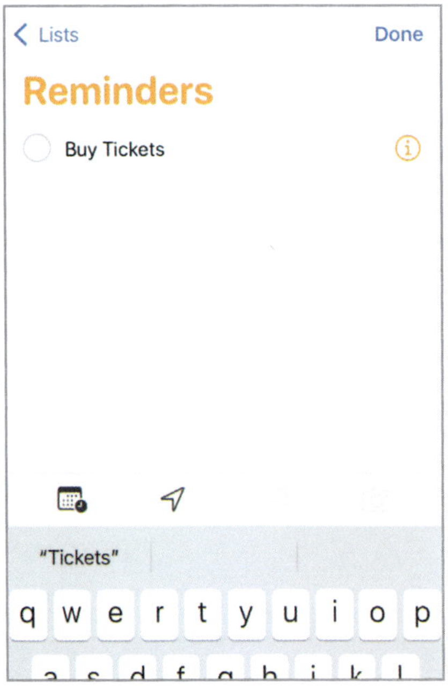

02 Add to the list

Type in your new reminder to make it appear in the list next to a check-list point and then tap on the Done button in the top-right corner of the screen when you're finished.

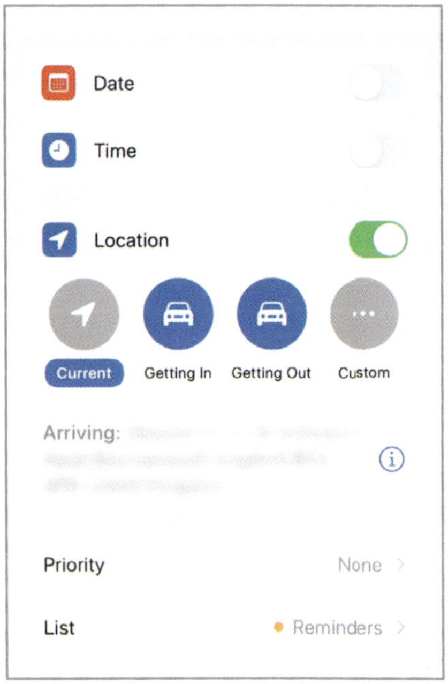

03 Turn it on

Tap on your reminder again and a 'i' icon will appear to the right of it. Tap it and you will bring up the task details screen and then tap the switch next to Location to 'On'.

Reminders in focus

Make more of the Reminders app

Arrivals and departures
The ability to set a location-based reminder which alerts you when arriving at or leaving a location is especially useful. It only takes a tap to set it up

Priorities
Priorities can be set for each task, but if you are using the app efficiently you will never need to manage your reminders at all. Let the app do the work for you

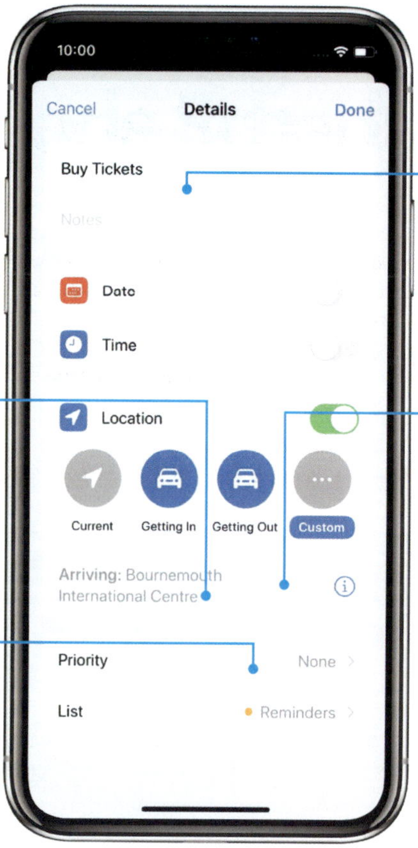

The description
Reminder details should be descriptive and to the point. When they alert, you will want to be able to quickly understand what you need to do wherever you are

Accurate locations
The full address, including postcode, will be shown so that you can be sure you have set up the right location. The map view is also a handy double-check

Guaranteed reminders
You can also set reminders to alert you on a specific date and time, which is the standard method, but by setting a location-based reminder with an extra alert based on time, you can be doubly sure that it will sound when you want it to. For very important tasks this will offer great peace of mind.

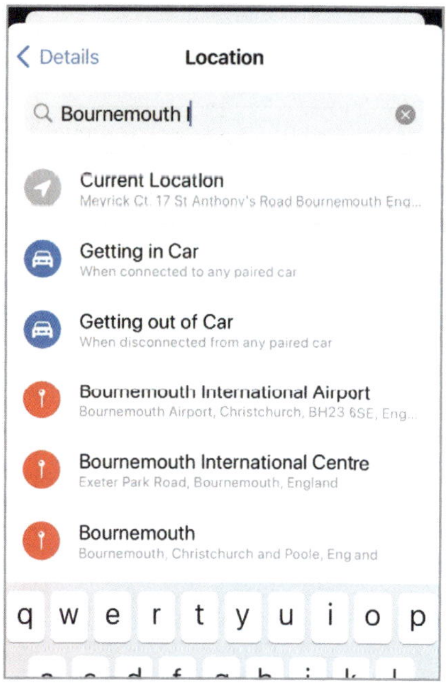

04 Choose the location
Tap on Custom and then type in an address for the Maps app to locate. When it has been found you will then be offered the option of viewing it on a map screen.

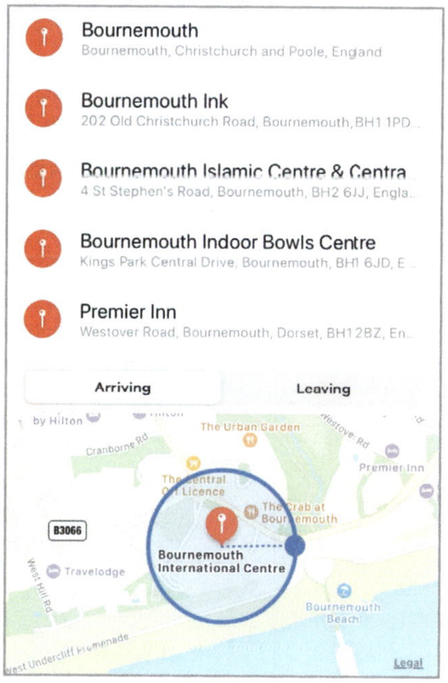

05 Double checking
You should use the map screen to ensure that the correct location has been found and you can then tap the top-left icon to return to the reminder details screen.

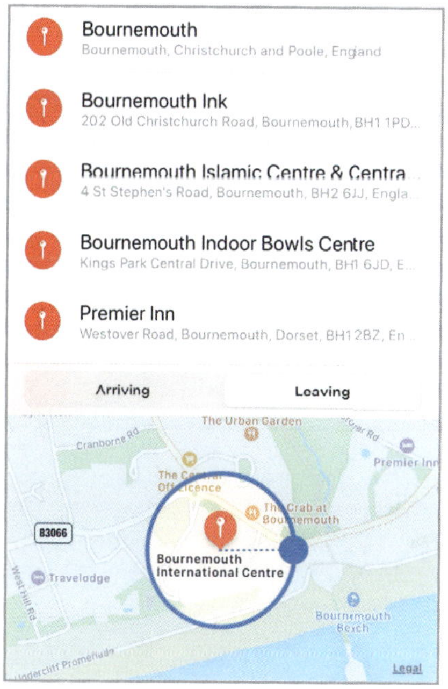

06 Leaving or arriving
You can now choose when the alert should sound – when you're arriving or leaving the location – and then tap Done. The reminder will only sound when you are there.

Keep track of current affairs with News

Only read the news that's relevant to your interests

There are countless third-party apps designed to help you curate online news for the topics and publications that you prefer, and each uses different methods to remain unique in a crowded market. Apple decided to join the party and offers the obviously titled News app to help us read only the news that is of interest to us or which is of great importance. It is all too easy to be overwhelmed with online content and being able to manage what you read will not only save time, but ensure that you get to read news that actually matters to you.

Apple has built a consistent interface that levels the various design styles of the different publications. The resultant app is exceptionally easy to use. Tap an article to read it and it will then be marked as read so you can move on to the next one. Recommended articles appear all the time, which means you always have something new to read. As you learn how the app works, Apple learns what you like and the content becomes even more personal and interesting as each day passes. It truly is the perfect example of less is more and is one of the very best mobile news readers available.

"It is all too easy to be overwhelmed with online content and being able to manage what you read will save time and ensure you get to read what matters"

News Read the world's news

01 View the articles
When you open the News app, you will be presented with all of the latest articles in a handy list relating to the relevant topic. Tap any article to read the content.

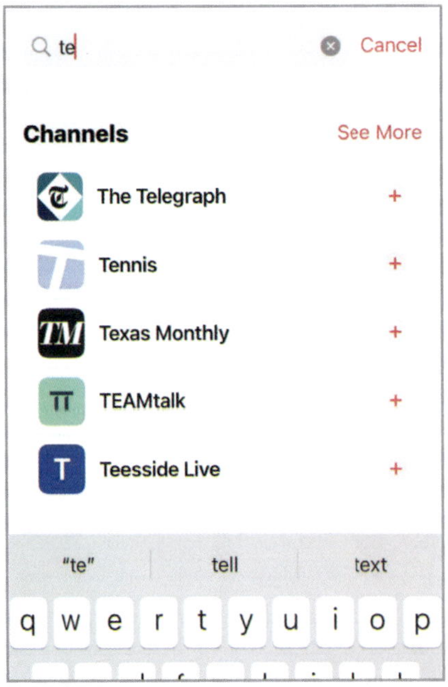

02 Searching
Tap on the Search bar in the Following section to look for a news channel, topic or story. As you start to type, suggestions will appear, grouped into categories.

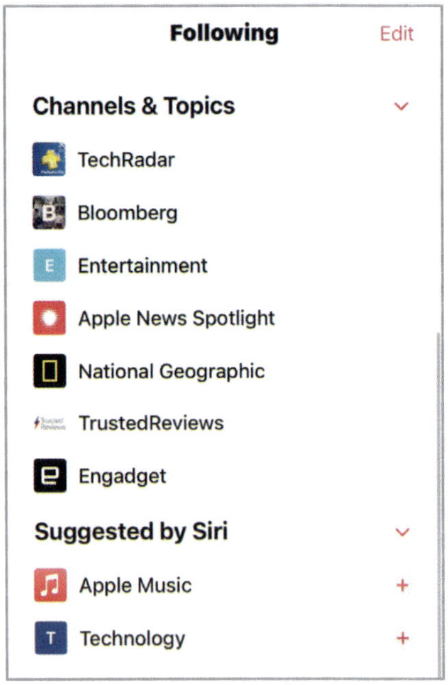

03 Suggestions
Go to Following at the bottom of the screen to choose what topics and channels you would like to see in your news feed. Any disliked channels can be managed through this tab also.

Reading your personalised news

What do all of the options and icons mean?

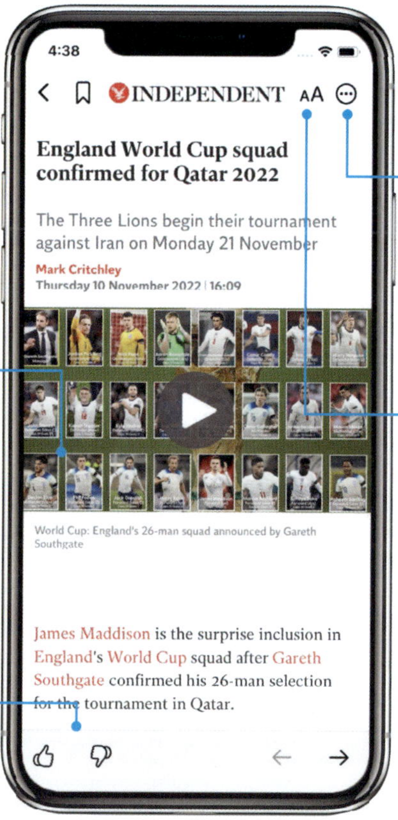

Sharing stories

By tapping the share icon you can instantly share the story using the various methods and apps, copy the text, add it to your Safari Reading List (for offline reading), open the story in Safari or mute the source

Love/dislike stories

You can cast a vote on all of the stories you read. If you 'love' a story then the app will source you more of that type of thing and if you rate to dislike it then expect to see less of that type of story

Saving stories

You can save your favourite stories by tapping on the share icon. To access them, go back out to the main menu, tap on 'Saved' and then use the tabs to view your saved stories and reading history, which is very handy

Change the font

By tapping here you will be able to beef up the size of the display font to make the news stories easier to read

Save and read anywhere

When you save an article in the News app, it will be stored in iCloud so that you can read it on any of your different devices. You can also read the articles offline, which is a useful feature when you do not have an internet connection available.

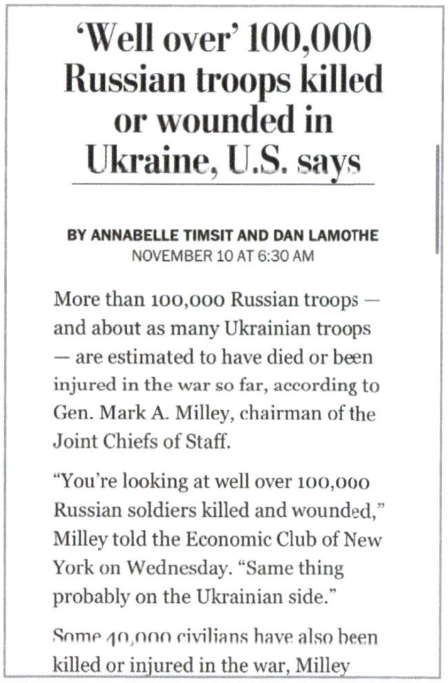

'Well over' 100,000 Russian troops killed or wounded in Ukraine, U.S. says

BY ANNABELLE TIMSIT AND DAN LAMOTHE
NOVEMBER 10 AT 6:30 AM

More than 100,000 Russian troops — and about as many Ukrainian troops — are estimated to have died or been injured in the war so far, according to Gen. Mark A. Milley, chairman of the Joint Chiefs of Staff.

"You're looking at well over 100,000 Russian soldiers killed and wounded," Milley told the Economic Club of New York on Wednesday. "Same thing probably on the Ukrainian side."

Some 40,000 civilians have also been killed or injured in the war, Milley

04 Near perfect presentation

You will immediately notice how well presented the articles are, no matter which publication they are from. They are consistently designed and extremely easy to read on the iPhone.

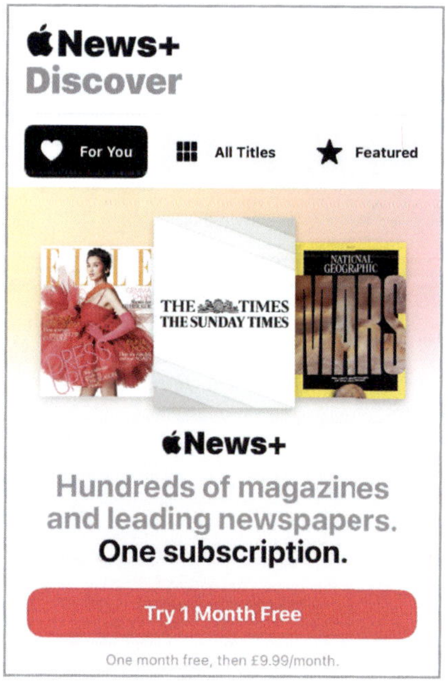

05 News+

For a monthly fee of £9.99/$9.99 you can access digital versions of over 150 print magazines and newspapers on your iPhone through News+. You can try it free for one month.

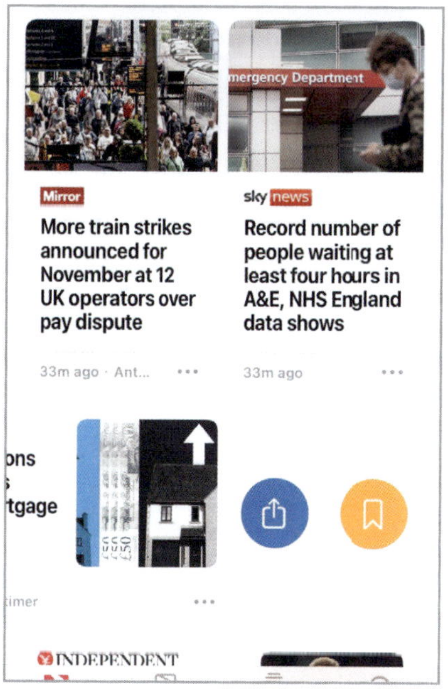

06 Share an article

When viewing your news feed, slide a story to the left and three options will appear. Love the story if you want to see more like it, Save to keep it for later or Share to post it on social media or via email.

Navigate your iPhone using Siri

Enjoy more voice control over your iPhone with a host of useful Siri commands

When it was introduced with the iPhone 4S, Siri seemed almost futuristic. Not only could you talk to your phone, but it would talk back and do things for you. Siri has stuck around but hasn't always been the most useful feature on our iPhones. This changed significantly when iOS 9 was released, making Siri a lot more versatile and responsive to your commands and the recent updates have placed more responsibility on Siri across numerous aspects of your device.

Siri can manage and neatly organise your photos, it can be used as a remote control for your Apple TV, and it can learn from the news stories you read to recommend words in the Quicktype bar. For example, if you search for a news story on Iceland, when you go to type in the Messages app, words relating to Iceland will be suggested in the Quicktype bar.

To call on Siri, hold down the Side button or simply say, "Hey Siri" (after first activating this feature in Settings>Siri & Search). If you don't say anything right away, examples of what you can ask will be listed. But trial and error works equally well and you will soon be surprised by how versatile Siri can be and how much smarter it is getting. After a while, you'll wonder how you ever managed without it.

Siri Go hands-free with your iPhone

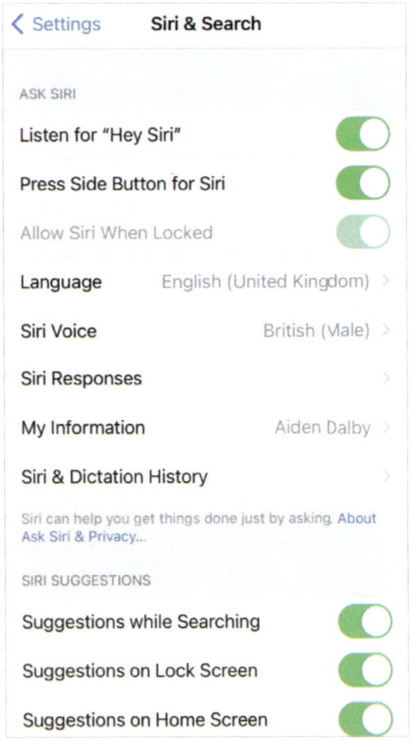

01 Change your settings

Hold down the Side button to summon Siri and then say, "Go to Siri Settings". On the Siri Settings screen, turn on the 'Allow "Hey Siri"' option. And then you can access Siri simply by saying that.

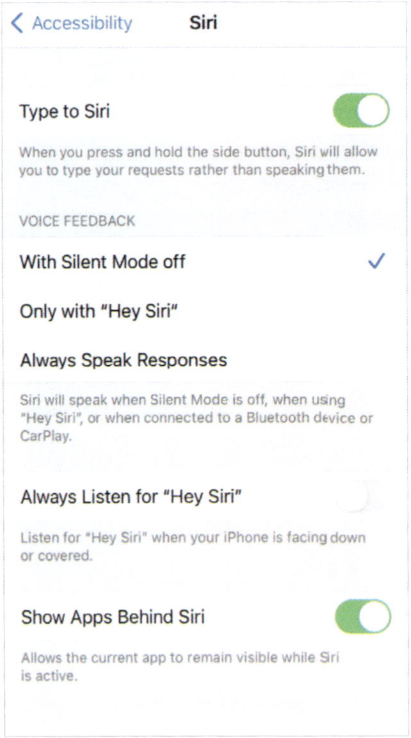

02 Type to Siri

In the settings app, go to General> Accessibility then scroll down to Siri. Here you can turn on Type to Siri. This is helpful when you want to use the AI in quiet places.

03 Get reminders

Whether it's for remembering to pick up something from the office or stop by at a particular location, you can ask Siri to set you a reminder. Just say "Remind me…" and Siri will fashion the reminder.

More Siri commands

Make the most of Siri's intelligence profile page options

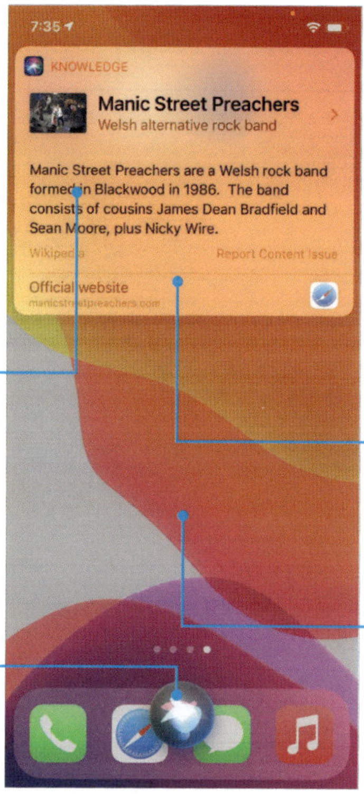

Responses
Whatever you ask of Siri, it will record your speech and analyse it to form a written representation of the question, think for a second, and then answer

Siri icon
After answering a question this icon will stay around for a few seconds before disappearing. Tap on it if you would like to ask it another question

Tap for more
Siri responses are usually concise but you can tap on them to see more information. By tapping on the response you can find links to websites with more information

Type to Siri
New to iOS 11, you can type your questions or commands. The keyboard does have a microphone option in the bottom-left corner so you get the best of both worlds

Teaching Siri
To make Siri even more practical, use phrases like, 'John Smith is my Dad' to teach relationships between people in your Contacts app and yourself. That way, you can say commands such as, 'FaceTime my Dad' to speed things up.

04 Siri's recommendations

Another cool feature, Siri will learn what you like based off your Safari history and searches then use this data to tailor content in other apps such as News to fit your taste.

05 Third-party app integration

Siri now works will a multitude of third-party apps, the sort which slot into your daily regime, such as WhatsApp for messaging or Facebook. You must give Siri permission for these tasks first.

06 The fun things

Remember, you can always ask fun things from Siri, such as to roll a die, pick a card or identify any songs that are played within earshot – Siri does have Shazam built in, after all.

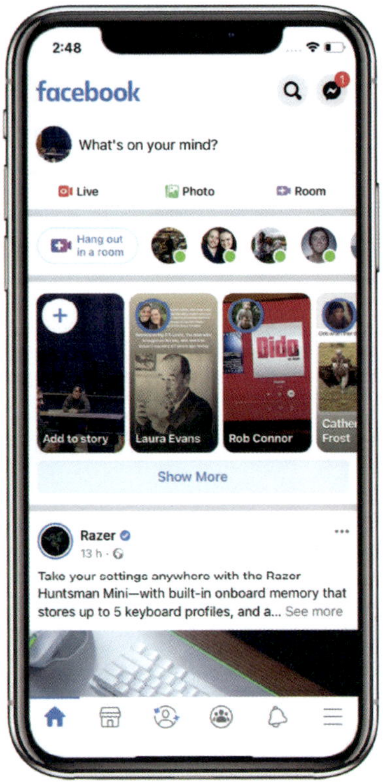

Learn to use Facebook

Discover why it's easier than ever to keep your friends and family in the loop on your social life

What was once an app integrated in iOS, Facebook is now something you must download from the App store, but you can still share content quickly and easily from within other apps. Interacting with the world's largest social network has never been easier. You can share a photo on Facebook straight from your Camera or Photos app, post your location using Maps, brag about high scores straight from games and, if you have your hands full, you can get Siri to do all the posting on your behalf.

You will first need to download the free Facebook app from the App Store and log into your account to access the Facebook features in your core apps. But once you're signed in, you'll be up and running and won't have to worry about logging in again.

With Facebook set up, you need not miss an important birthday or event ever again because all Facebook events are integrated straight into your Calendar app. What's more, all of your friends' profile information can also be integrated straight into your Contacts app, so if your friends update an email address or phone number, then you will automatically stay bang up to date. This is very useful indeed.

There are many more features to the Facebook app than we will cover here, so after you've used our tips, try exploring the app. Events, groups and more can all be created within the app and shared with friends. With Facebook, it has never been easier to keep your friends and family updated. Here we show you how to get started.

Facebook Get set up and update your status

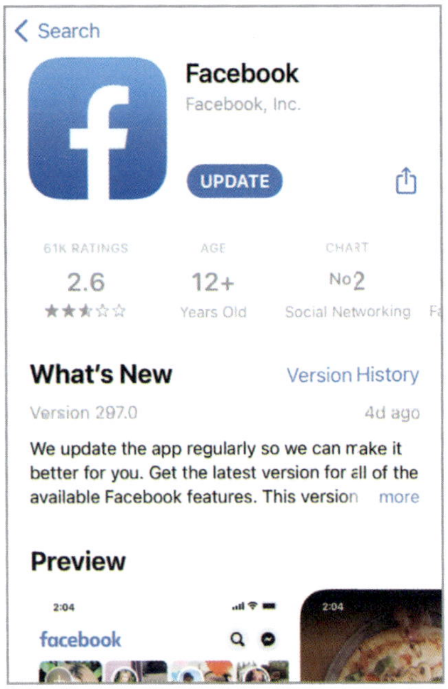

01 Setting up

Download the free Facebook app from the App Store and then load the app once it has installed. Log in using your account details if you have an account or create one if you don't.

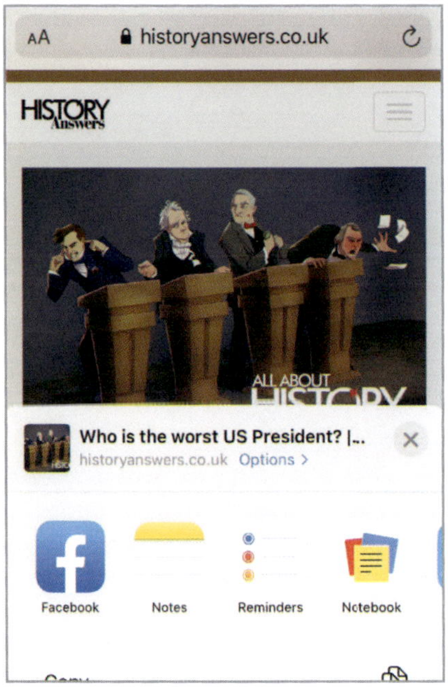

02 Sharing through apps

You can now tap the Share button while in any other app and scroll right to select the Facebook option. This allows you to share content without having to open the standalone Facebook app.

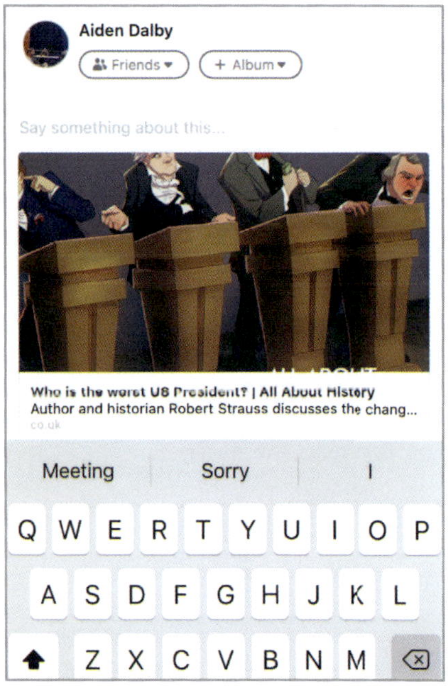

03 Add details

Tapping the Facebook share option brings up a window with the desired content attached. You can now amend the text, add a location and tailor the people who see the post. Tap Post to finish.

Share a post using Facebook

How to create different types of posts

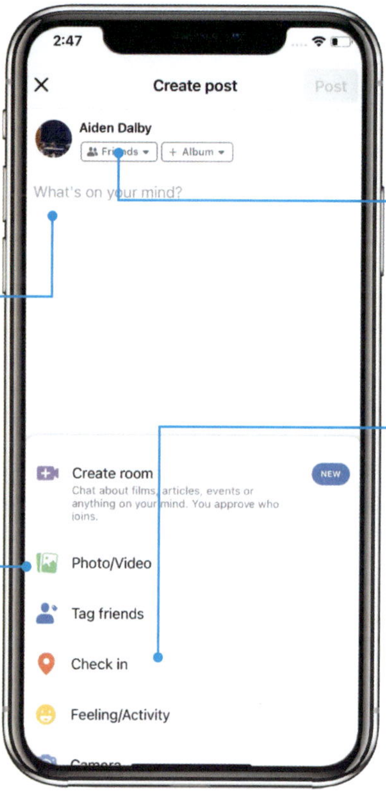

Write a post

Facebook has a character limit of 63,206, so don't worry about running out of space. If you want to delete the post before it goes online, tap on Cancel in the top-left corner

Share a photo

You can post photos to Facebook using the Photos app, or you can tap here to use the app's tools for sharing. It may just come down to using whichever tools you prefer

Tag a friend

By tapping on this you can include any friends you have on Facebook into the post. This means that the post will appear on their wall to and their friends can see it

Add a location

If you have location services enabled, then when you tap here, Facebook will recommend places based on where you are. If not, you can use the search tool to look for locations

Your Story

At the top of your Facebook feed will be your Facebook Stories. These are posts that appear for 24 hours before being permanently deleted. You can tap the '+' icon on this section to create your own Story post using images, text and even music.

04 Uploading pictures

With Facebook integration, this couldn't be easier. In your Photos app, go to the image you wish to upload and tap the share button. Now choose Facebook and add details before posting it.

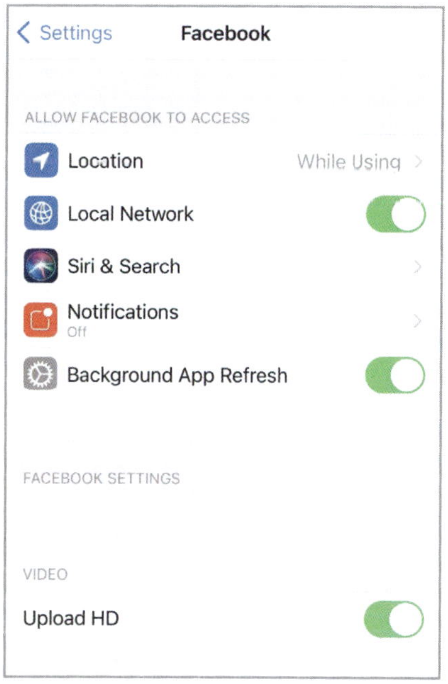

05 Facebook settings

In the Settings app, tap on Facebook to see a range of options that relate to Facebook integration on your iPhone, such as Notifications and Background App Refresh.

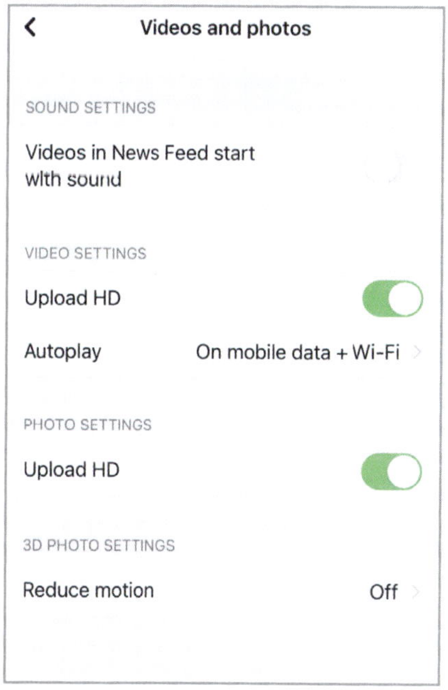

06 Other settings

There are also settings found while in the Facebook app. Tap on the bottom-right hand icon and scroll down until you find Settings. By taping on Videos and Photos you can turn off Auto-Play.

Get to grips with the X app

X can be integrated into iOS, so we'll show you how to share your thoughts and links from within your favourite apps

There's little doubt that X (formally Twitter) is a social phenomenon. This simple application allows you to share thoughts – or rather 'posts' – with people around the world almost as soon as they enter your head. Likewise, you'll see what's on the minds of people that you choose to follow.

The advent of smartphones has taken this concept to the next level by allowing you to post at any time, wherever you are in the world, making it easier than ever to speak your mind. Now, the posting process is even easier with your iPhone. X was once integrated seamlessly into Apple's operating system; but now you must download it from the App store. You no longer need to open the X app and find a photo to upload or copy links from your web browser; simply do it directly from within an app. It's so easy that your X activity will increase once you get to grips with it.

In this tutorial, we guide you through setting up the X app on your iPhone, and how to post from your favourite apps.

> "You no longer need to open a X app, find a photo to upload or copy links from your web browser; simply do it from within the app"

X Get the most from the app

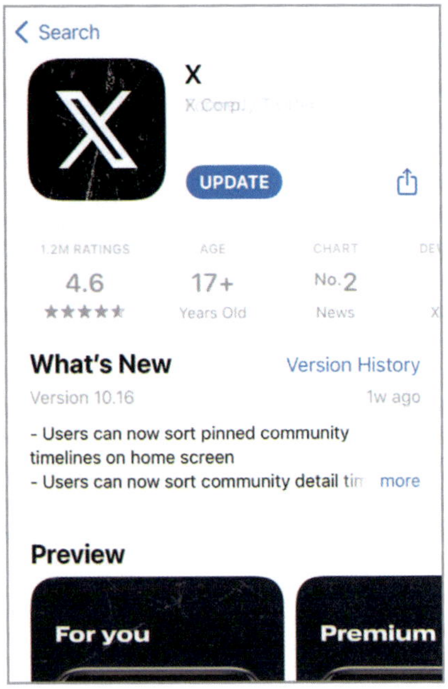

01 Setting up

Download the free X app from the App Store and then load it up when it has installed. Log in using your account details or create an account and you'll be ready to go.

02 Sharing through apps

You can now tap the share button while in any other app and select the X option. This allows you to share content quickly and easily without having to open the standalone X app.

03 Adding details

Tapping the X share option brings up a window with the desired content attached. You can now add text and a location before tapping Post to share the post with others.

Master sending a post

Post text, images and more from your iPhone X

Go live
Instead of posting, you can start a live video stream. Before you go live you can add a brief description and check the camera's view before going live

X polls
Tap this icon to post a survey. Your question must be under 280 characters and you can have up to four choices of answer

Character limit
One of the interesting things about X is all posts have to be 280 characters or less. The remaining character count is represented by this circle. It will fill up as you type and when it's full, you're out of characters. The overflow will be marked red

Locations
Tap this icon to add a location to your posts. To enable location services go into the Settings app and select 'While using'

Camera comments
You can also post from within the Camera app, allowing you to upload, comment and caption your pictures almost as soon as they are captured. Keeping others informed on your life has never been so easy.

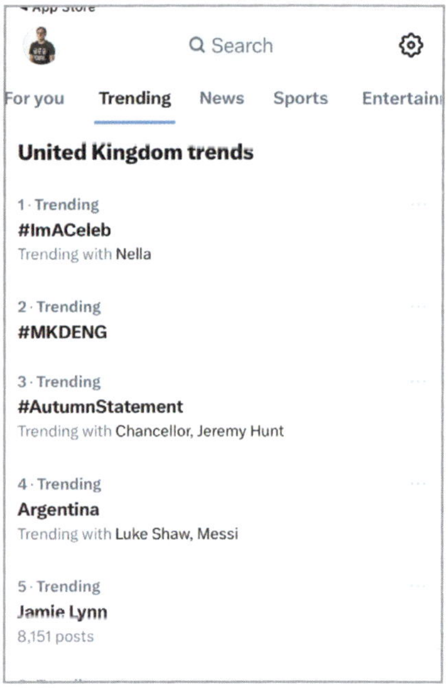

04 Check what's trending

Tap on the magnifying glass to go to the Search tab and see what stories are currently trending. These trends will also be tailored depending on who you follow or what you post about.

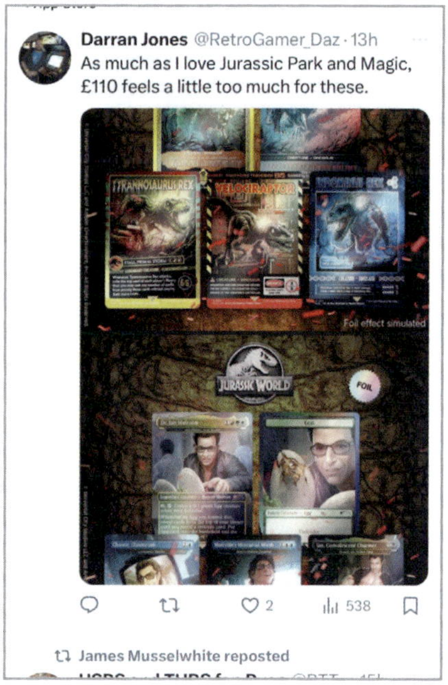

05 Your timeline

By default, X will show the recommended posts first but by tapping on the icon in the top-right corner you can change this so post are displayed in the order they were sent.

06 Notifications

You will receive notifications if someone mentions you, posts at you or when you are quoted. Highlights are also displayed here - these are posts you missed but may be of interest to you.

Track your fitness

By feeding health, nutrition and fitness data into a single dashboard, Apple's Health app can help you stay in peak condition

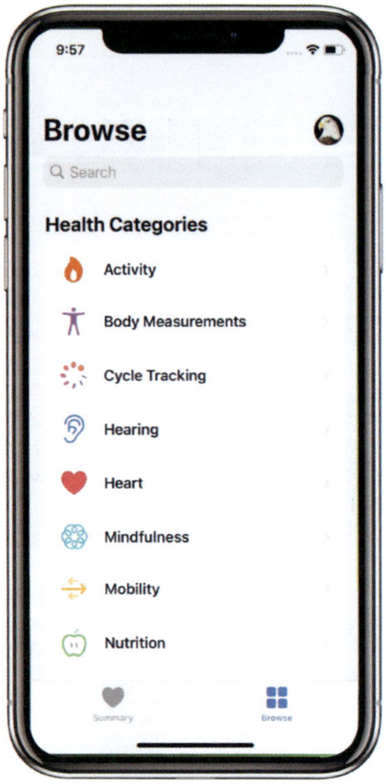

One of the biggest growth areas within the smartphone and tablet market over the past few years has been all things health. Not only is the App Store awash with health-related offerings but there are also many gadgets that work in tandem with your iDevice. All of these generate data to help build up a picture of your overall health, but the problem has always been that they operate in isolation. Until now, that is.

With Health you can link your various apps and devices, collecting all of the data about your fitness in one place. It means, for instance, that information from a gadget charting the number of steps you take each day can be combined with the data gathered by an app measuring your heart rate. When taken in tandem you get a much better overview than if you were looking at the two sets of data separately.

Many apps and gadgets can work and communicate with Health, and the same goes for the Apple Watch, which is also compatible with the app. In this tutorial, we will look at how you can get the most out of the Health app and how to indicate exactly what data can and cannot be shared. You can also rest easily knowing that your data is safe since Apple promises that all information on a locked phone is encrypted and that it will only be shared if you want it to.

Health Get started with Health

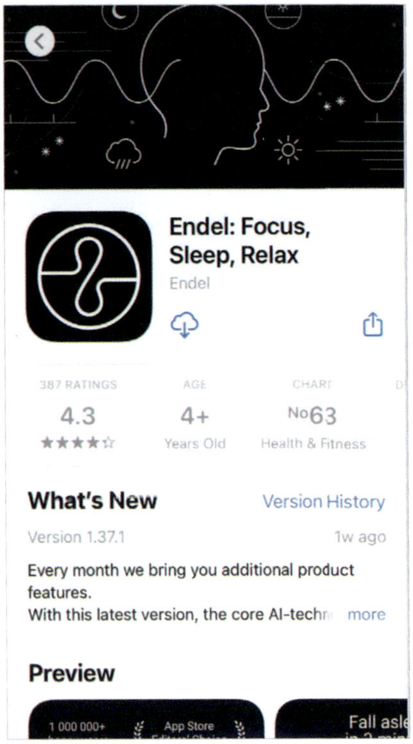

01 Install an app

Health is a central hub which collects data from your other health apps. Install apps from the App Store, reading the description carefully to ensure compatibility with Health.

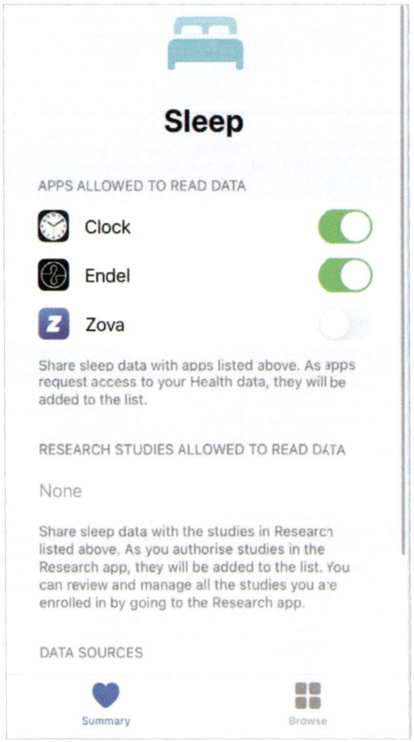

02 Open Health app

After a compatible app has installed, open Health and go to Activity, Sleep, or Steps. Scroll down to Data Sources & Access and you will see your compatible apps.

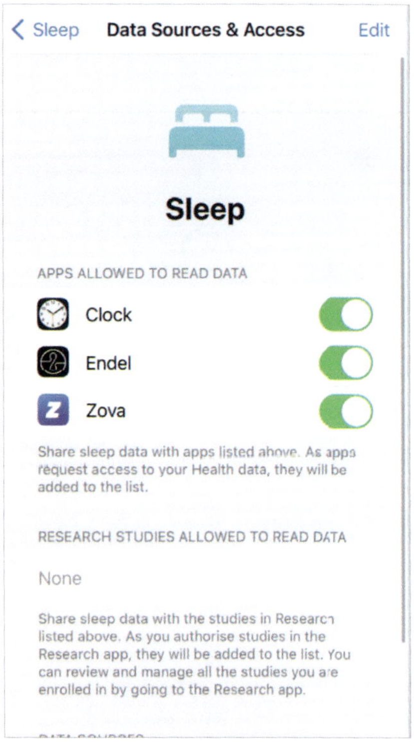

03 Allow apps access

An app will ask for permission to send data to Health. By moving the slider to green, you are indicating you wish to share a particular set of data, such as active calories.

Keep track of your information

Create your Medical ID

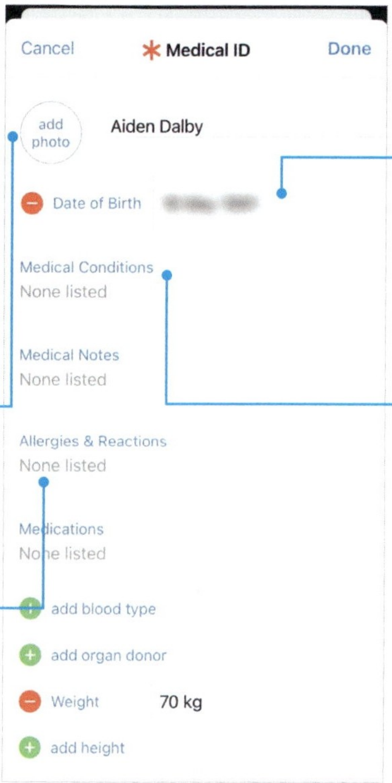

Name and picture

The medical card is personal to you. It asks you to upload a profile image, give your name and your date of birth, details that are vital for medical staff in helping to identify you should you need emergency treatment

Notes and allergies

There is space for specific medical instructions under the notes category and you can also jot down any allergies and reactions you may have. As you work your way through the card, you will be asked for contact details of family too

Show when locked

By tapping the Medical ID tab in Health, you can create an emergency health card. With the slider next to Show When Locked moved to green, the card will be accessible on locked iPhones – perfect for assisting medical experts in a hurry

Medical Conditions

Note down any medical conditions that are known to you and perhaps ask your doctor if you are unsure or would like further guidance. The more details you include here, the more useful your emergency card can become

Apple's Healthkit

Third-party developers are able to work with the Health app thanks to Apple's Healthkit. The beauty of this is that you are able to purchase any fitness gadgets you wish and have them work together with your iPhone, ensuring you can tailor and control how your health is monitored.

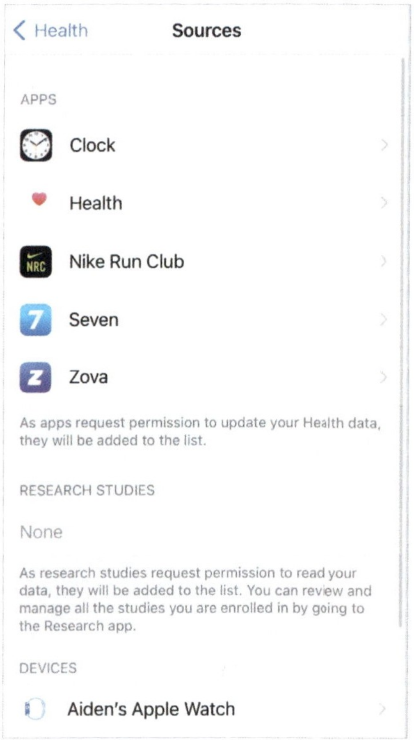

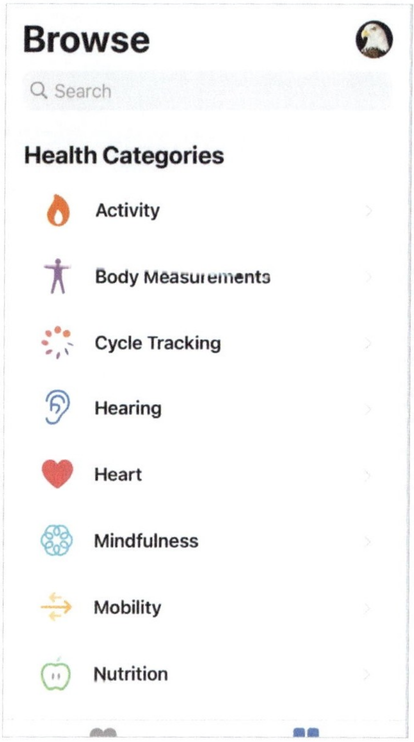

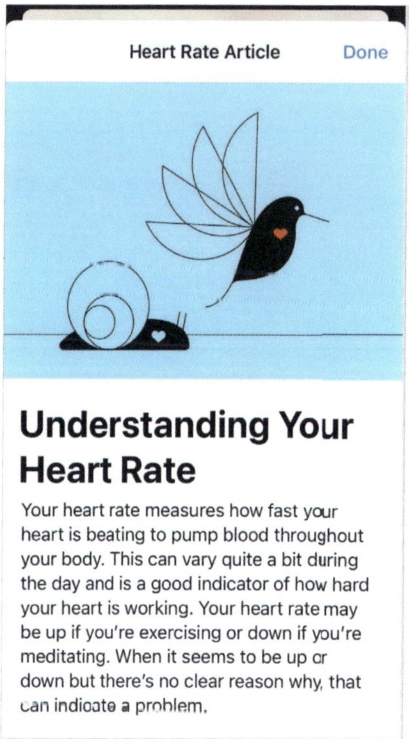

04 Reading data

You can also change app permissions by going into the Settings app, tapping on Health and then Data Access & Devices. Tap on an app and then select what permissions you wan to allow.

05 Check health data

As you use those third-party apps, data will be written to Health. Tapping Browse lets you see a break down of the results, letting you monitor activity, nutrition, sleep and more.

06 Learn more

By scrolling down the Summary page you can find articles meant to help you learn more about your health. These include how to sleep better, and understanding heart rates.

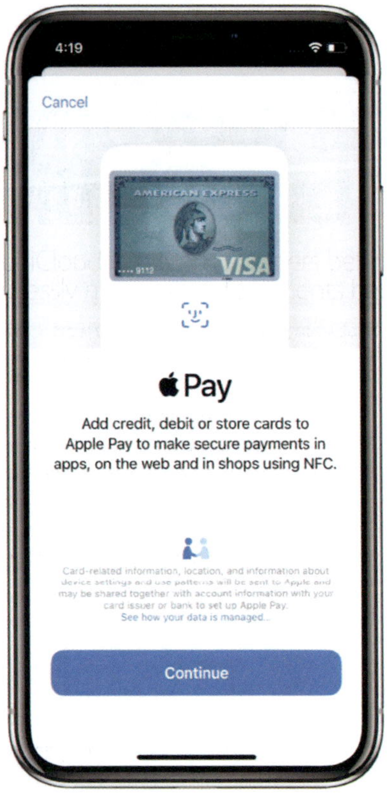

Set up and start using Apple Pay

The innovative Apple Pay makes the payment process intuitive and easy

The introduction of Apple Pay marked a huge change in the way we pay for goods. It is exceptionally convenient, highly secure and after a few uses you may find it tiresome to use a traditional card to pay for things.

The retailer will not receive any information about you when you make a purchase and a unique code is used for every single transaction for added security. The use of Face ID is a major security benefit and with the extremely natural Wallet app managing all of your payment cards, there really are no downsides to using Apple Pay. We will show you how to set up a debit or credit card to use with Apple Pay and also demonstrate how to make a payment, which is easier than you may expect.

Indeed, when you first use Apple Pay, it doesn't even feel as though you have bought anything, but trust us when we say that it will soon become the most natural thing in the world. The Wallet app has the potential to replace much of what you currently carry in your real wallet and makes your iPhone more useful than ever before in emergencies or during normal daily use. The future of payments is already here.

> "With the extremely natural Wallet app managing all of your payment cards, there really are no downsides to using Apple Pay"

Wallet Add new payment cards

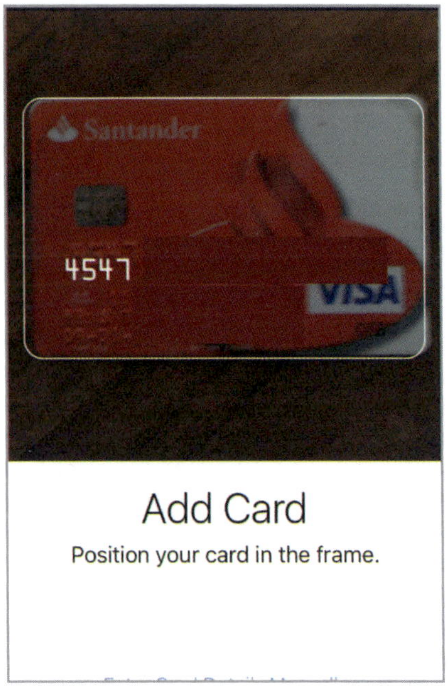

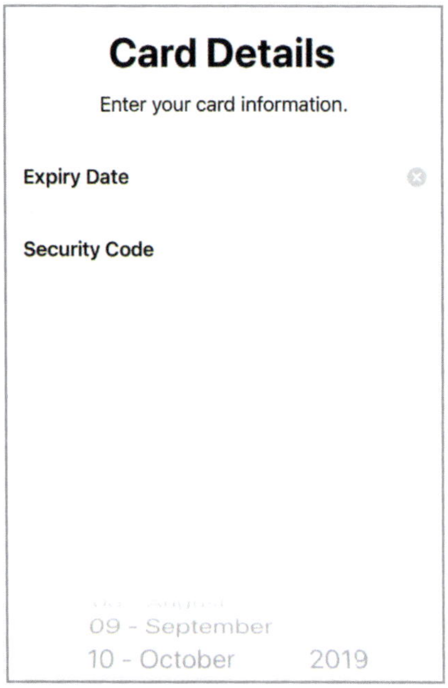

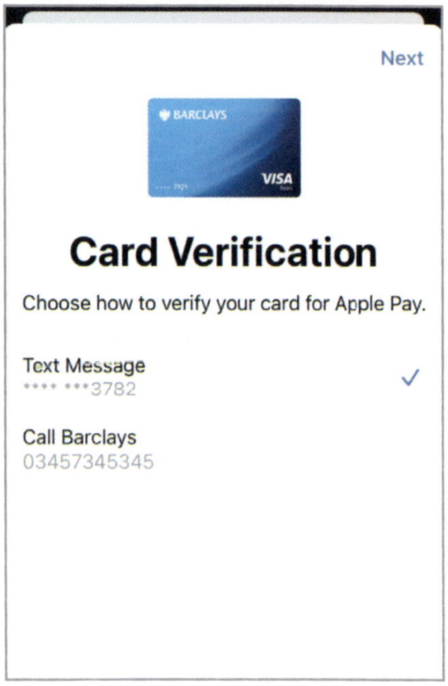

01 Scan a card

In the Wallet app, tap the '+' icon. If you haven't already, you must set up Face ID and a Passcode. Once you've done this, point your iPhone at a payment card and it will immediately recognise it.

02 Verify your card

In order to verify your details, you now have to navigate through a selection of screens to input information such as the security code to validate your payment card.

03 The final verification

Your bank will require you to call them, or you can opt to receive a text verification code instead. The entire process only takes a couple of minutes and is very simple.

Make payments securely and flexibly

Take a look at the Wallet app's interface

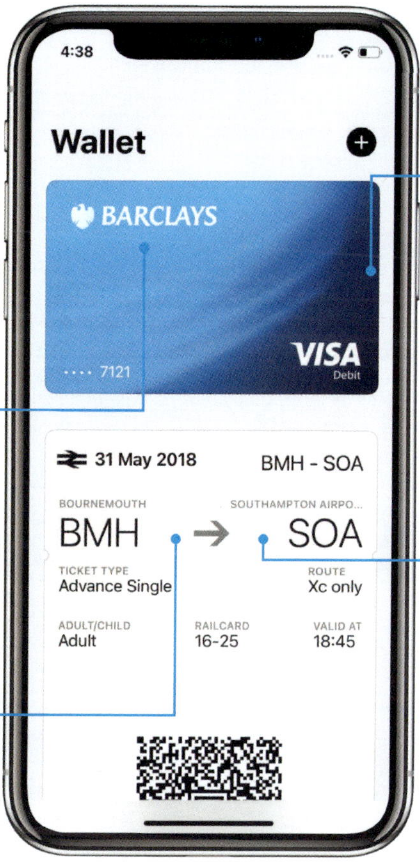

Use it anywhere
You should be able to use Apple Pay in almost any retail location that supports wireless payments. Conveniently, it now offers almost universal coverage

It's very secure
The setup process is efficient, but still extremely secure so you need not worry about fraudulent payments occurring through the reliable Apple Pay system

Multiple cards
You can set up any debit or credit card that supports the Apple Pay system. The range of banks that support it is growing

Other Wallet uses
Don't forget that Wallet can also be used for loyalty cards and online tickets if the establishments support the Wallet service

Apple Pay on Apple Watch
If you choose to set up Apple Pay to make payments from an Apple Watch, you will need to ensure that the iPhone is close to it when making payments. Almost all of the security verification is done on the iPhone for obvious reasons.

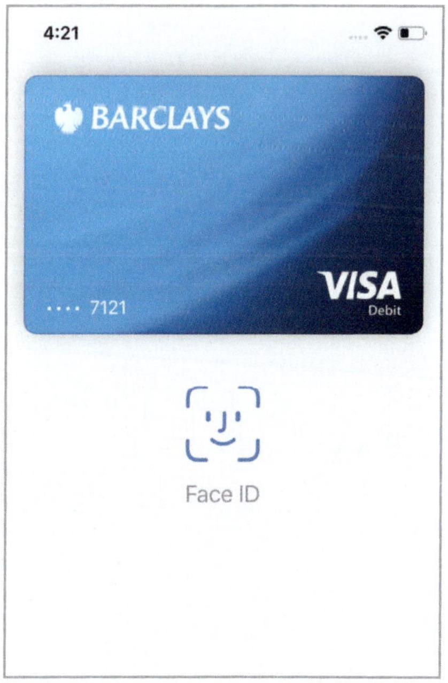

04 Time to pay

You can quickly bring up Apple Pay by tapping on the side button. You will see the above screen and you will need to confirm your identity using Face ID.

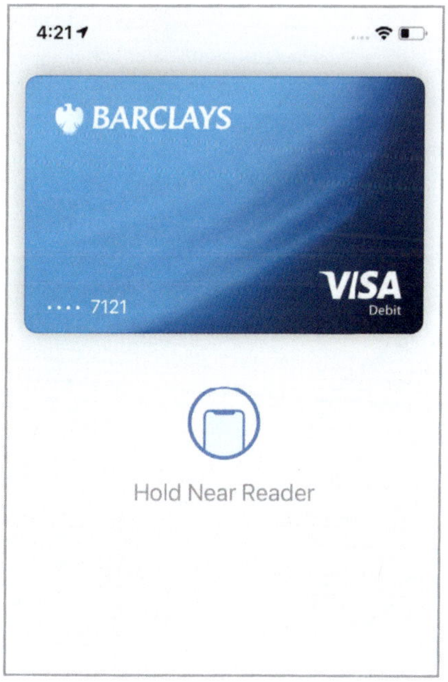

05 Find the reader

Once you've confirmed your identity, simply hold your phone near the payment reader in the store and the payment should be made in an instant, just like that.

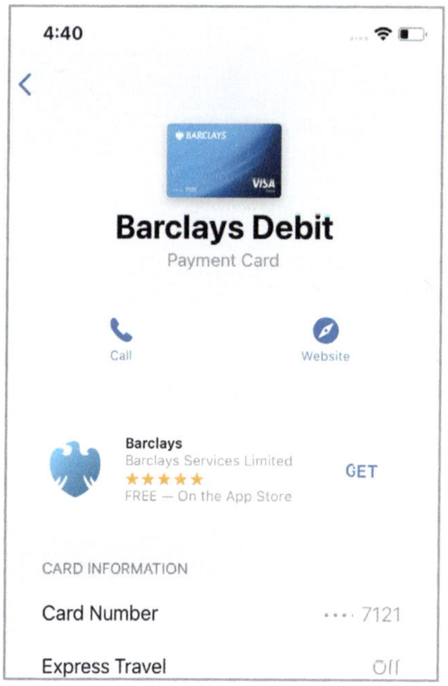

06 Check your payments

All your payments will be listed in the Wallet app below your card so that you can check for accuracy. If anything is wrong, tap on the more icon to bring up contact information for your bank.

The next step

"Once you've mastered the basics and move on to the next level, there will be no stopping you!"

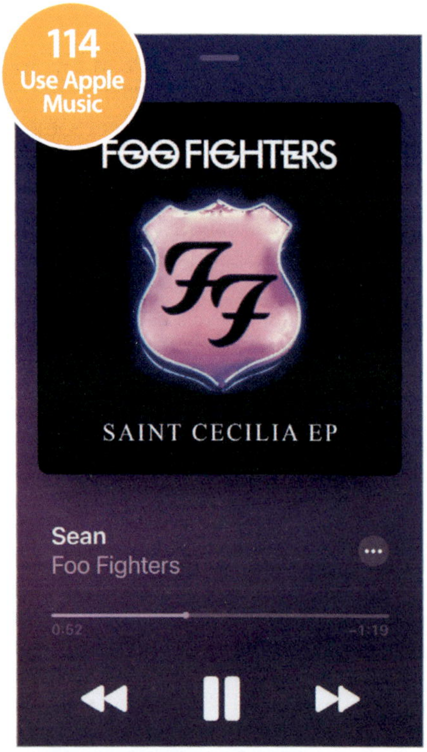

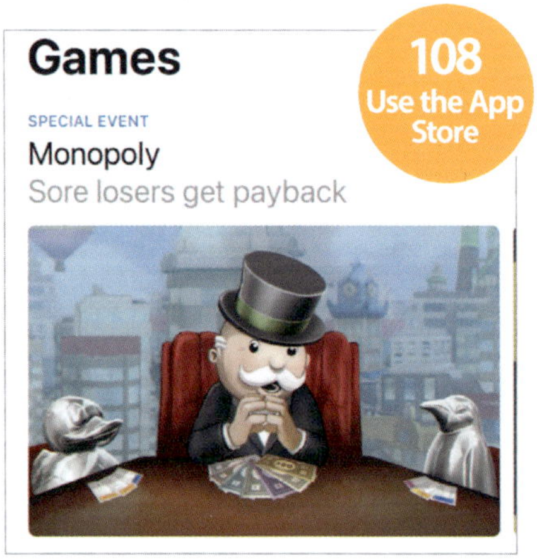

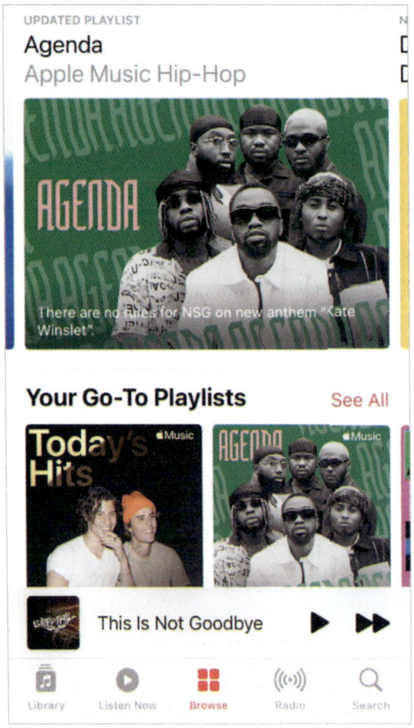

Start downloading powerful apps and multimedia

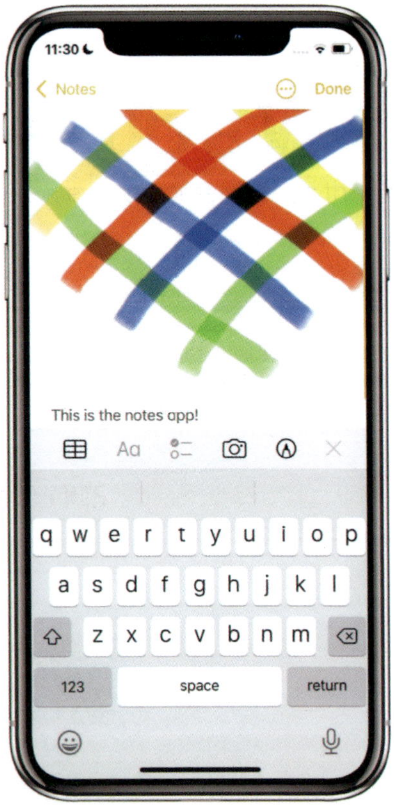

Jot down your thoughts in Notes

The latest iteration of the Notes app embraces customisation so you can create more visual and interactive notes

Once upon a time the Notes app was capable of simply recording text, but that all changed with the more recent releases of iOS. The app is now capable of offering all the media-related features you could need, and it is more connected to other apps than ever before.

Besides typing in a thought or reminder to yourself, you can now make it more visual by inserting images and adding sketches, which can be drawn with a variety of drawing tools. Each object can be moved around freely within the note and all of the data is held in place if/when you want to share your note with others. Also compatible are web links, photos and task lists, so you can save everything you need. The main listing interface is as simple as ever and the sorting is logical, with full support for the global searching mechanism within iOS, meaning it can work with so many other apps.

Apple has taken the basic Notes app and added just the right number of carefully created features to make it capable of managing many aspects of your life with ease – it's a long way from how it used to be, but the updated version should cater for your every need.

"Apple has taken the basic Notes app and added just the right number of carefully created new features to make it capable of managing many aspects of your life"

Notes Make interactive notes

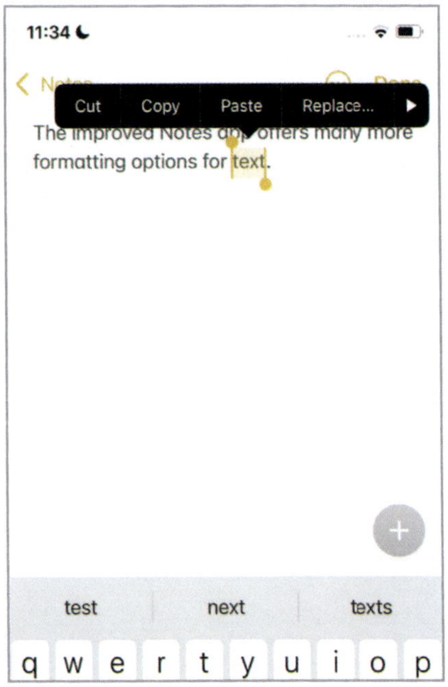

01 Type and format your note

Open Notes and type as you wish. When complete, tap and hold some text in the new Notes app to see formatting options. You can change the text to bold, italic or underlined. You can also indent it.

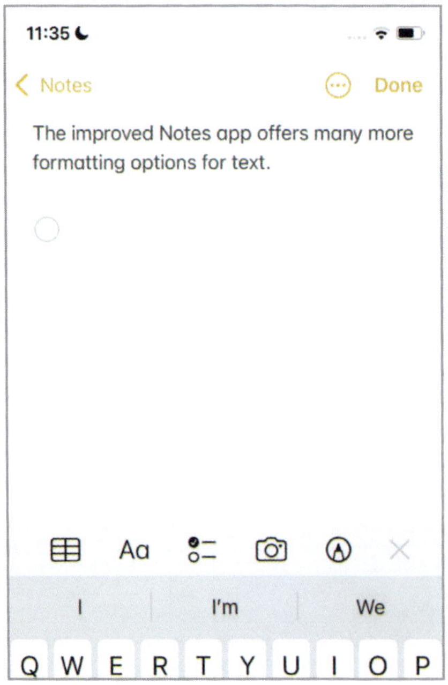

02 A bulleted list

To create a bullet point list, tap the small 'Aa' option at the bottom of the screen, just above the keyboard. You will see the various listing options that you can include in a new note you're creating.

03 Sketches and rulers

If you want to illustrate your note with a doodle, tap the sketching icon to see all of the drawing tools available. Then just start sketching! Your doodle will be inserted into the note immediately.

Design some dynamic notes

Better understand the note-taking interface

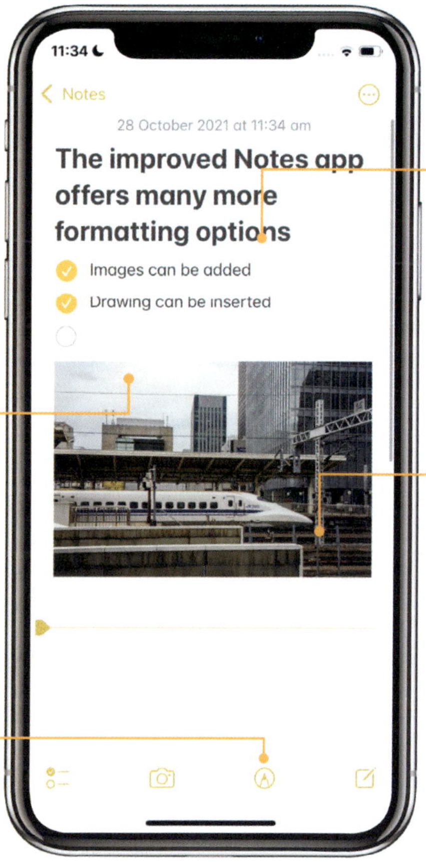

Logical lists
You can now create bulleted and numbered lists as well as task lists. Multiple text formatting options are also available when selecting text

Insert images
The ability to insert images in all notes adds much more personality and potentially makes the entire experience more flexible

Move around
Tap and hold a sketch or photo to move it to another part of the note. You can choose any order for your notes

Simple sketches
A variety of sketching tools are included so that you can draw, write or sketch quick notes to review later. They add a personal touch

Upgrade your iCloud account
When you first try to share a web object or other item to the Notes app you may see a message asking you to 'go to Notes and upgrade your iCloud account'. The best way to do this is to turn off Notes in the iCloud settings screen and then turn them on again.

04 Insert images

Tap the camera icon to select or capture an image to insert into your note. Portrait photos take up the most space, but they do look neater within the space.

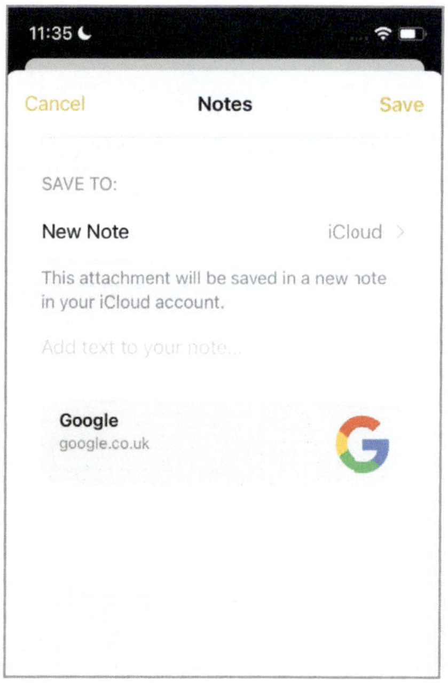

05 Share to Notes

You can share data to Notes from other apps and a new note will be automatically created. This includes web links and many other items that are worth saving.

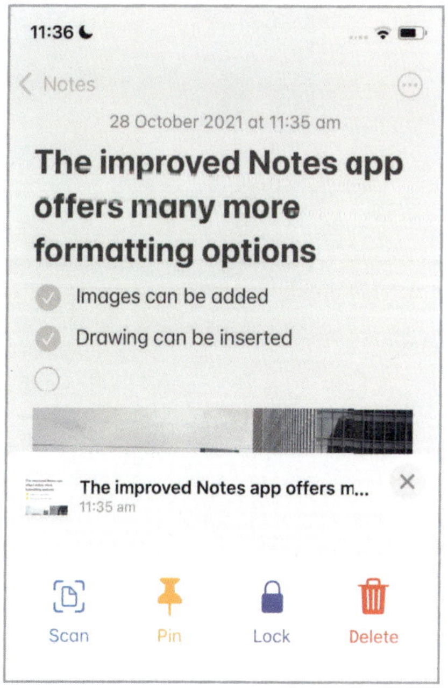

06 Share from Notes

All of the formatting, images and sketches will stay in place when you share a note through another app. This makes the process more worthwhile than ever before.

Use the App Store

One of the best things about your iPhone is that it can be upgraded on the go. The App Store allows you to make the iPhone even better than it already is

The iPhone is magical, and a major part of the magic is that you can expand its capabilities with cheap or even free apps. The App Store has been a roaring success for Apple, with over 1.8 million applications to choose from – not to mention the $155 billion that it has made for developers.

In iOS 11, the App Store got a major redesign, with the store being split into two major sections; Apps and Games. There is also the Today section, which has interviews with developers of the most popular apps in the store, and let's not forget Apple Arcade.

The software tends to be incredibly good value – it's amazing what your iPhone can do after just a 99p investment. And navigating the App Store is easy too. Simply use the section icons at the bottom of the interface to jump to the various parts of the store, or see what's popular in the Search tab. Follow our guide on these pages to get the most from the App Store.

"The number and range of applications on offer is stunning"

Featured apps
The Featured section of the each page showcases the latest releases or updates and the most popular downloads

Options
From the bottom menu you can see the featured apps in the Today section, search for a specific app or see which ones need to be updated

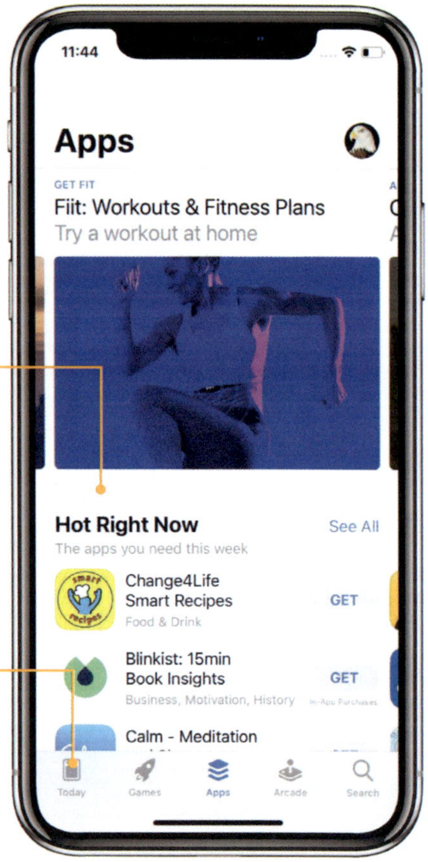

App Store Learn how to use the App Store

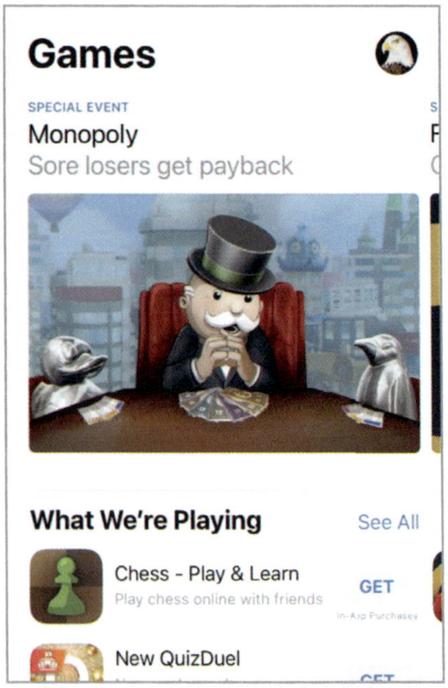

01 Discover great games

At the top of this section will be the featured games - these can be new titles or ones with major updates. Scrolling down will take you to see the most popular downloads.

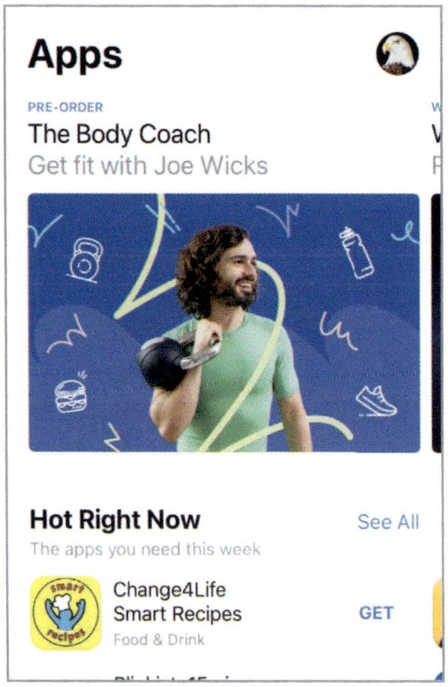

02 Find new apps

The App section is formatted the same way as the Games section. Scroll down to the very bottom to find links to redeem a gift card or download code, as well as a parent's guide to the store.

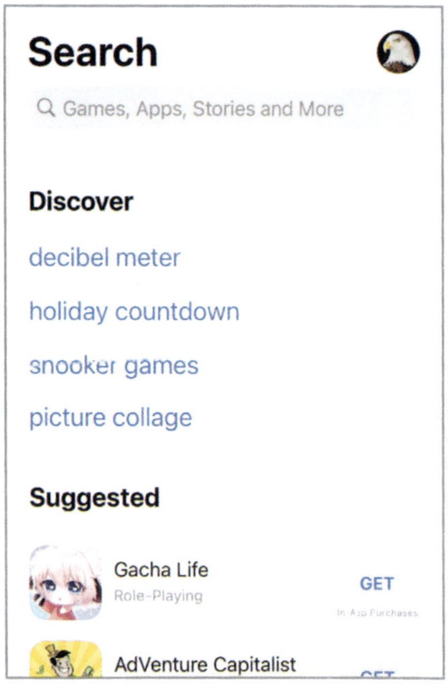

03 Search for specifics

If you know the name of the app that you would like to download, then all you need to do is tap the Search tab along the bottom and type keywords into the blank field.

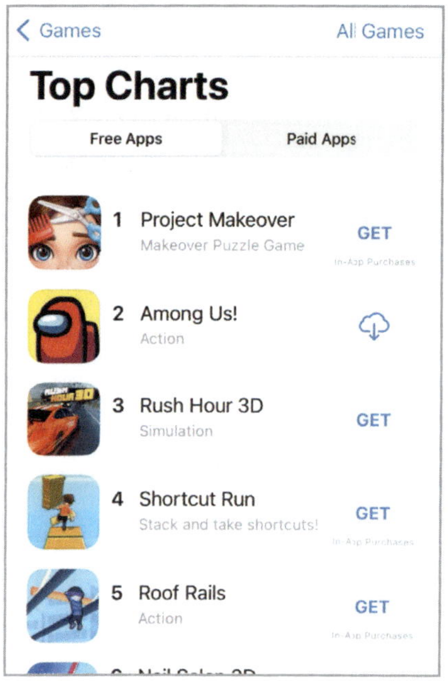

04 Browse the charts

Apple allows you to see the Top Paid and Top Free charts for both Apps and Games, which is rather useful. Scroll down to the Top Charts in either section and press See All.

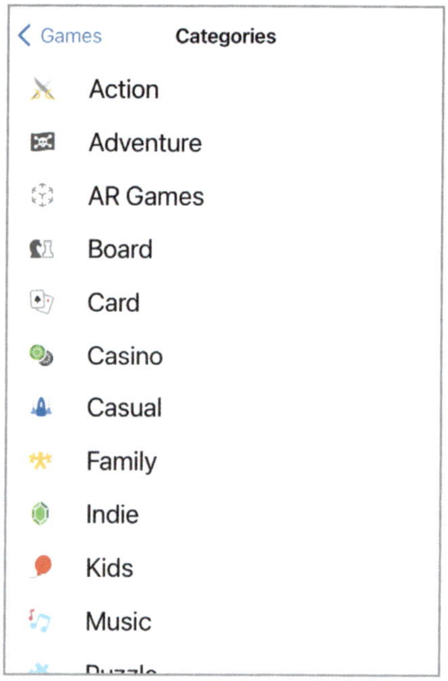

05 Check out categories

If you are not completely sure what you want, but have a general idea, then you can scroll down to the Categories section of either the App or Games tab and tap on See All.

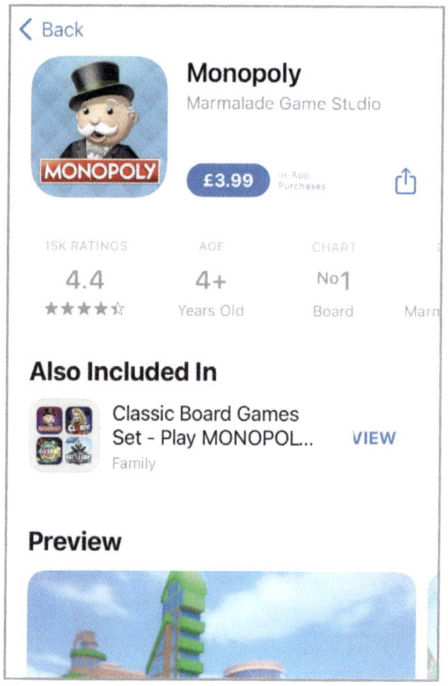

06 Buy an app

When you are ready to buy an app, the process is very straightforward. Simply tap on the price, and a small menu will appear. Tap Purchase, and then enter your Apple ID password.

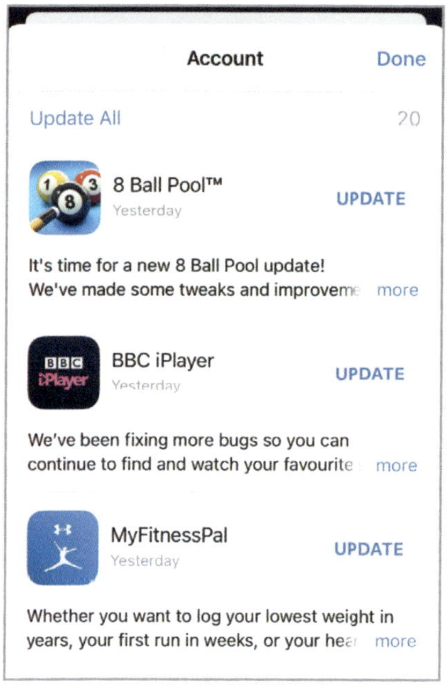

07 Update an app

Over time, apps get updated. Tap on the profile icon at the top of any page, and see if you have any that require updating. The screen tells you if all of your apps are up to date.

08 Explore the Today section

The Today section is the first thing you will see when you open the app. There is a daily group of apps and games with descriptions on why you may want them, as well as a game and app of the day.

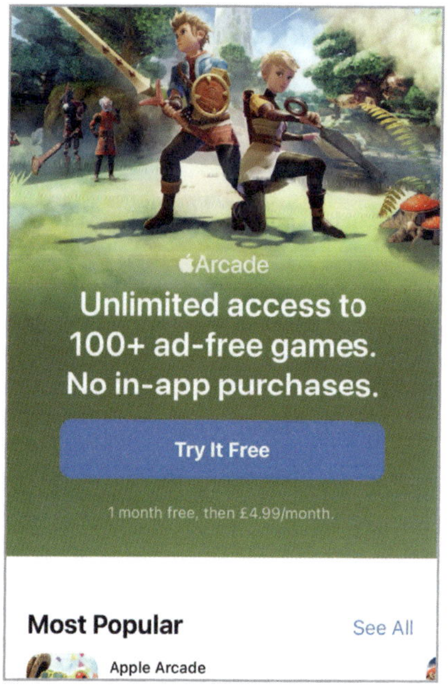

09 Apple Arcade

For a monthly fee you can access Apple's exclusive library of games. The best part of these games is they are all ad-free and there are no in-game purchases. You can try it free for one month.

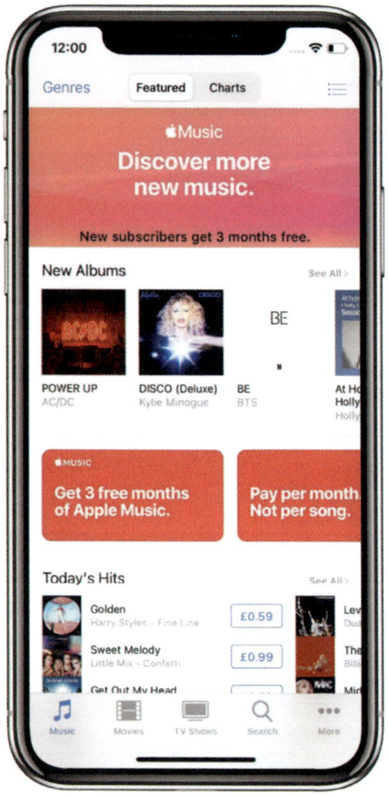

Get the most from iTunes

You can download all your favourite movies and music from iTunes without connecting your iPhone to your computer

Being able to download music and movies on the go is one of the best things about Apple's devices, and this is especially true of the iPhone. With the enormous selection available on the iTunes Store, you're unlikely to be lost for something to suit your mood.

It's not just music and movies though, as you can also download TV programmes, audiobooks and ringtones too. Thanks to the intelligent Siri service, you can even hold your phone up to a sound source, ask Siri "What's the song?" and after listening intently for a few seconds, Siri will not only name the song and artist, but will also allow you to buy the track straight from iTunes – genius!

All of your iTunes purchases are automatically backed-up to your iCloud, meaning that you can get them beamed straight to all devices running the same account (go to Settings>Music and toggle the Automatic Downloads to 'On') and re-download any past purchases by tapping on the 'More' tab within the iTunes Store app, going to the Purchased section and tapping the cloud icon next to any item not on your iPhone.

To use iTunes on the iPhone you will need an account to get going, but once that has been set up you can buy songs, apps or movies wherever you are. All you need is a credit card (or, failing that, some iTunes gift cards).

"With the enormous selection of music available to download, you are unlikely to be lost for something to suit your mood"

iTunes Take advantage of iTunes

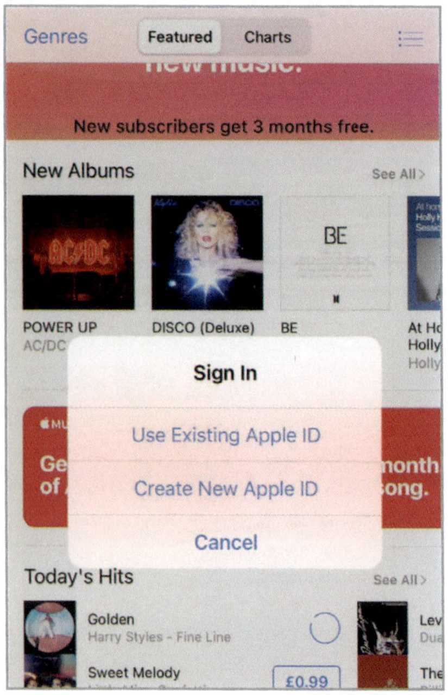

01 Buy music, movies and more

If you would like to make purchases you will need to log in or create a new account. Tap on an item in order to buy it and choose Create Apple ID or Use Existing Apple ID.

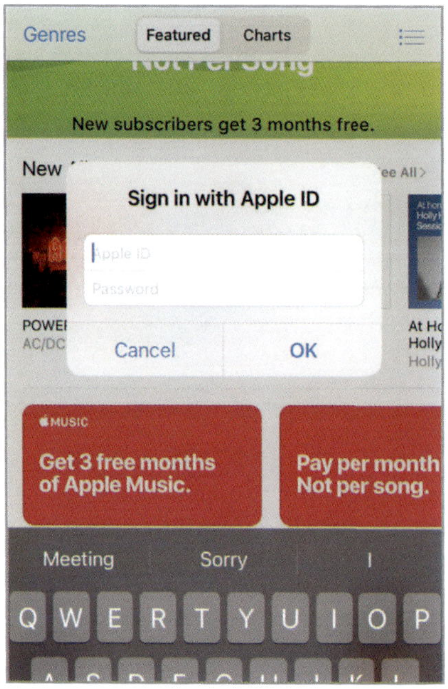

02 Agree to the terms

Once you have signed in or signed up and confirmed your location, you will then be offered the terms and conditions for using the account, which you must agree to if you want to continue.

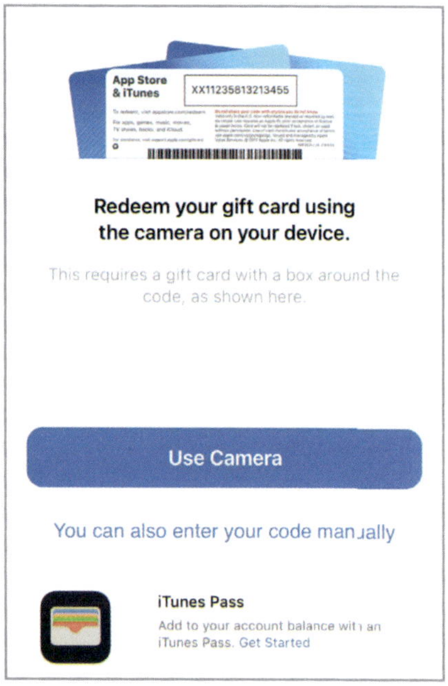

03 Enter your details

You will need to provide your credit card details at this stage. If you do not want to enter your details, then simply buy an iTunes gift card and enter the number here.

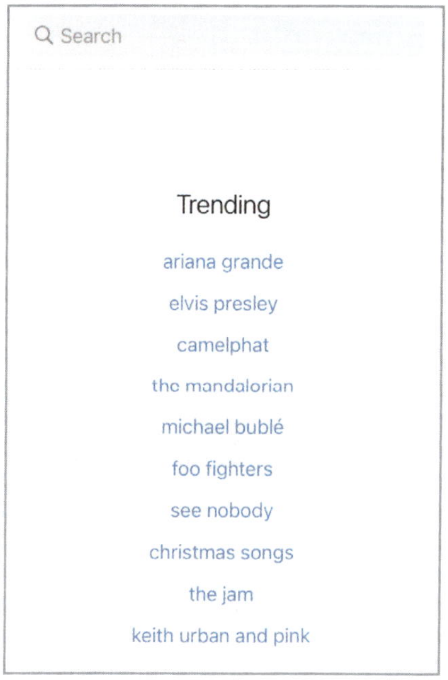

04 Search

To search for something, all you need to do is tap the Search tab at the bottom. It will offer suggestions as you type. If you spot the one you are after, tap it.

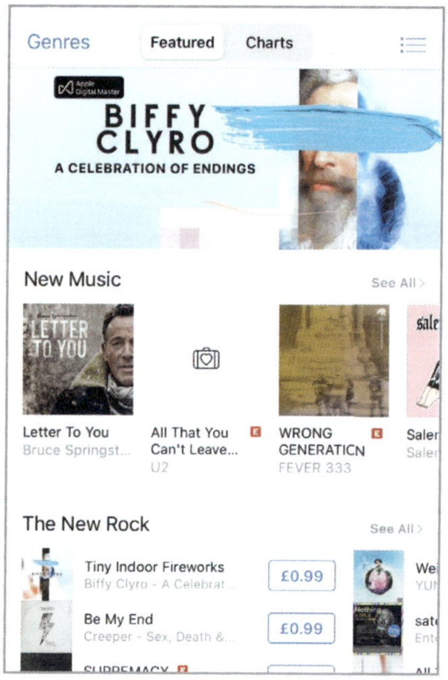

05 Music

Tap on the Music tab that sits along the bottom-left corner of the screen in order to see new releases, recommended tracks, and some truly classic albums to download.

06 Films

Now tap on the Movies tab, which is also along the bottom of the screen, in order to search for a movie that you feel like watching, or see what has just been released.

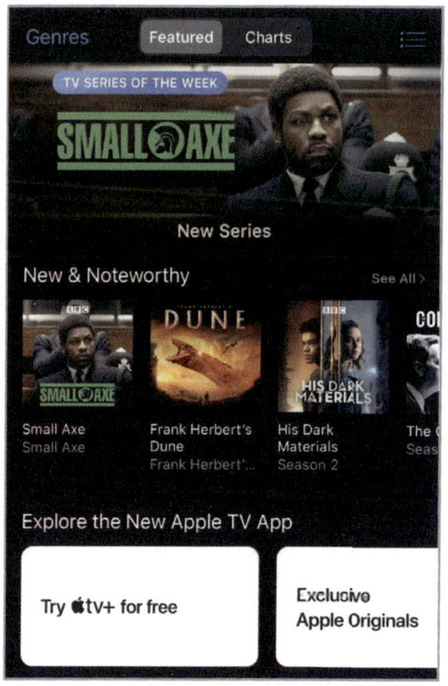

07 TV Programmes

The iTunes interface makes it incredibly easy to download single episodes of TV shows. Tap on the TV Shows tab, find a show and make a purchase there and then.

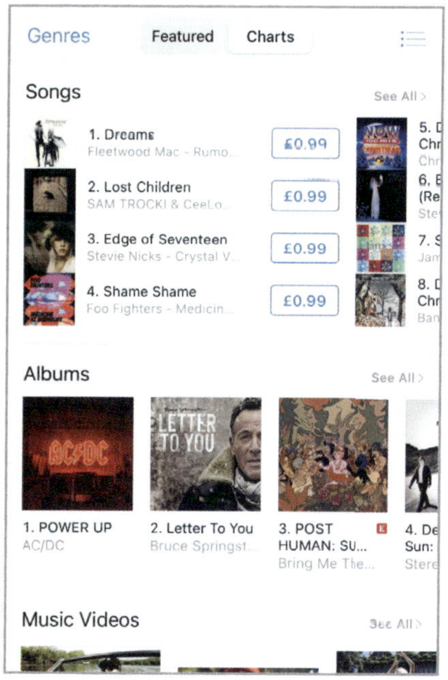

08 Charts

If you would like to see what is popular right now, you are able to view charts for music, films and television programmes. You might even find a new favourite.

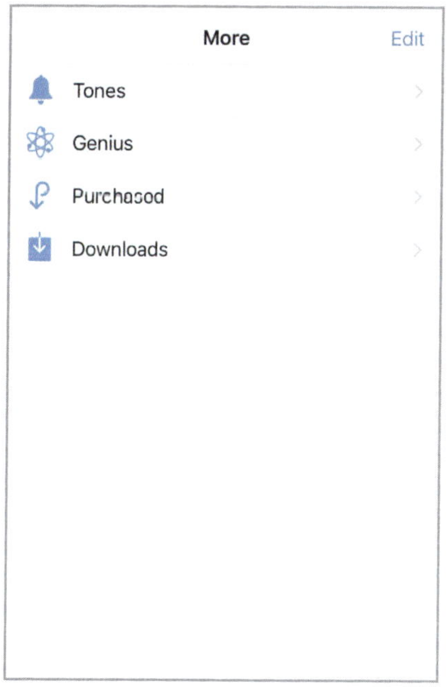

09 Get more

The More tab brings up a selection of extra options, which include purchasing ringtones, viewing already-made purchases or viewing anything that you are currently downloading.

Buy and view TV shows and films

With iTunes on your iPhone, you can download and watch single episodes or whole TV series

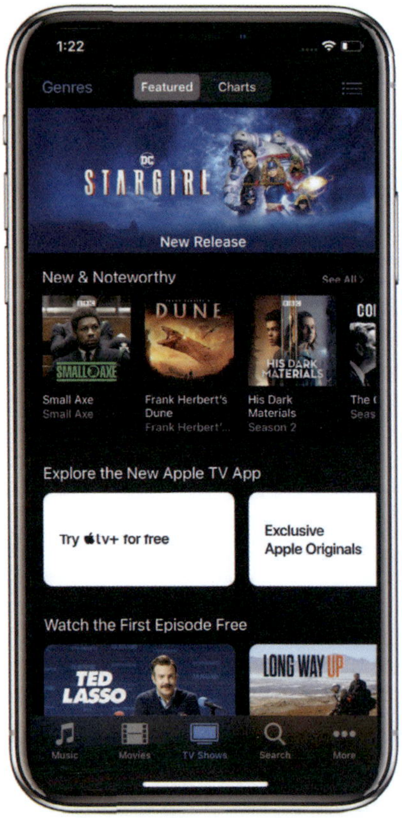

There are some really excellent movies and television programmes on offer in the iTunes Store, there are even whole series with up to two dozen episodes available. You can easily watch old favourites again or catch up with programmes and series that you might have missed. The typical price of a television programme is just £1.89, but a series depends on the number of episodes that are contained within it, of course. It is always cheaper to buy a series than buying episodes individually. Sometimes the price of a film can be as little as £1.99.

The iTunes Store makes it really easy to find what you want and you can see featured content, top charts, the latest releases and so on. There are informative descriptions and useful reviews from others who have purchased the film or show. The video is downloaded to your phone and you can watch it at anytime and anywhere. The video quality on the new iPhone 12 is fantastic thanks to its high-resolution screen, and iTunes even sells 4K films to take advantage of it. Or if you don't fancy owning the film, you can always just rent it for a limited amount of time.

"The video is downloaded to the phone and you can watch it at anytime and anywhere – even standing at the bus stop"

iTunes Store Download top TV programmes or films

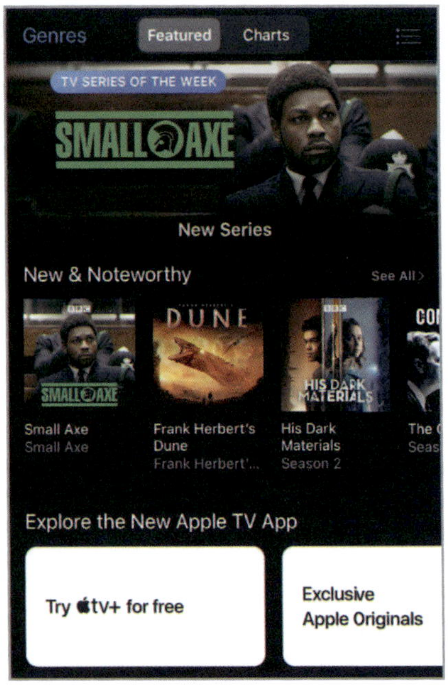

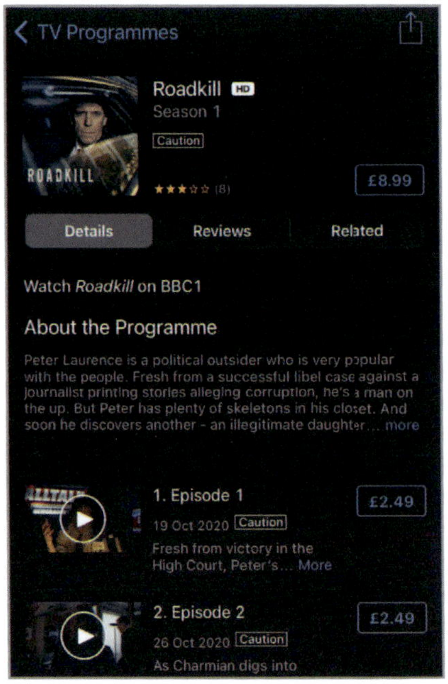

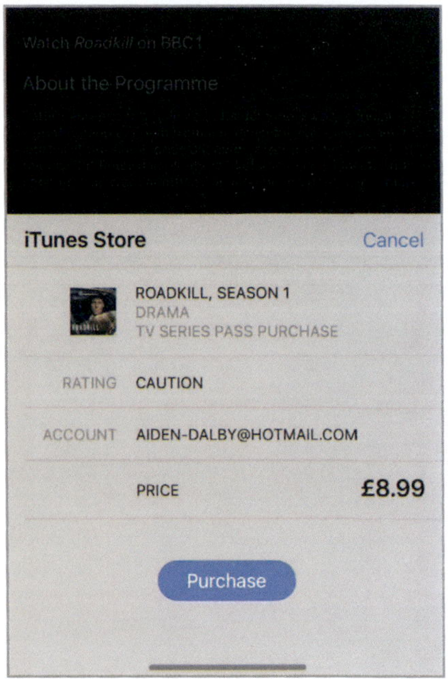

01 Browse the store

Tap iTunes on the iPhone's Home screen to start the app and then select TV Shows or Movies at the bottom. The screen is packed with links to the latest or most popular content. Browse the store.

02 Select a programme

When you find one you like, select it to see a description or use the Search tab if you know what you want. With a series, each episode's plot is described. Tap the price to download it.

03 Pay for it

After pressing the price button, a pop-up menu will appear. If you tap Purchase you will be prompted to enter your password to confirm the sale. or you can double-tap the Side button to use Apple Pay.

Browse TV programmes

Fire up iTunes to browse the TV programmes and see reviews and related items

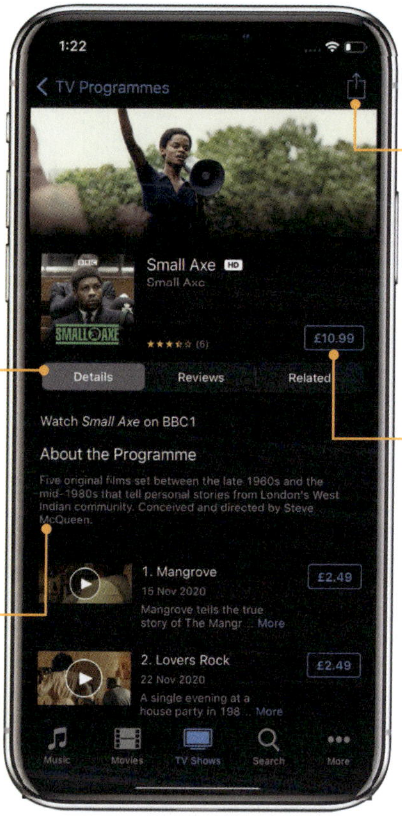

Browse the tabs
There are three tabs that are jam-packed with useful information, and Details describes the programme or series. Click Reviews in order to see what others think and Related to see similar programmes

Read the description
There are brief details about the programme or each of the episodes of a series. Tap here in order to view the full description before you tap the Buy button

Share your finds
Why not share programmes you like? Click this button to email a link, send a message, post it on Facebook or Twitter, or even buy it as a gift for someone

Buy it now
This button shows the price. In this case, it is the whole season, hence the bigger price tag. Tap this to begin downloading it. Keep in mind the space required

Films
As well as a great selection of new and classic TV series, iTunes also provides you with a fantastic selection of movies. Tap the Films tab at the bottom of the screen. As with TV Programmes, films are sorted into genres to help you search.

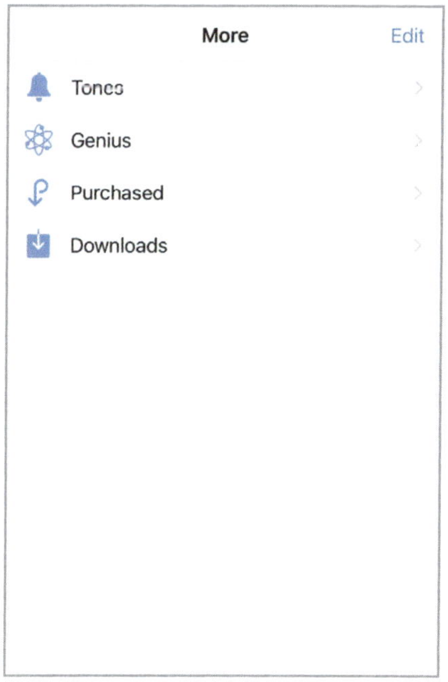

04 Check the download

Press the More button and then Downloads to see the episode or series of episodes downloading. Feel free to use the Pause button and continue later if it takes too long.

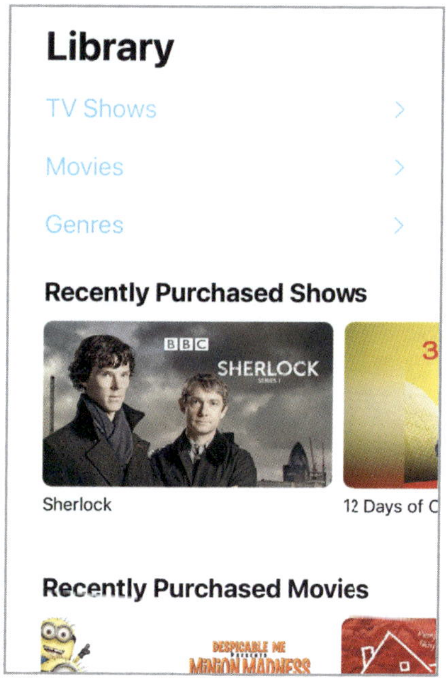

05 Go to Videos

The TV shows and films that are downloaded are added to the Apple TV app. Tap on its icon to see your purchases. Tap the Films or TV Programmes tab and then select the video.

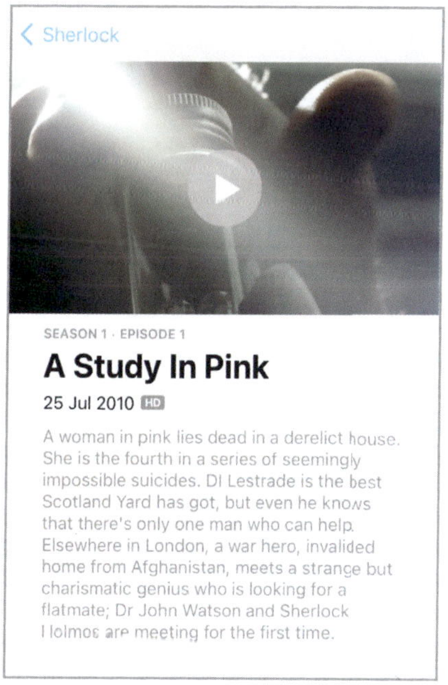

06 Watch the video

Tap a video, then the Play button to start it. Tap the screen for various controls. The TV programme can be paused and stopped, and then restarted to watch it again.

Explore Apple Music

How to enjoy Apple's revolutionary Music app and streaming service

Whereas previously the Music app was merely a way to enjoy listening to your iTunes music collection on the move, it has grown into a much bigger beast. While you can still enjoy your music through your iPhone, the focus of the app has now shifted to the Apple Music service, which lets you discover new music and stream it straight to your device for a monthly subscription fee. You can enjoy this service for free for three months to try it out, but if you don't fancy continuing with the service you'll have to turn off automatic renewal to prevent being charged for the service after the trial period.

You can also use the Browse section of the Apple Music app to see what the current hot music is as well as listen to meticulously-compiled playlists from music experts. Apple also has three 24/7 radio stations; Music1, Music Hits and Music Country, that can be accessed through the Music app. If you're into music and open to trying new tunes, the Music app has come of age for you.

Stream music
The Apple Music service lets you stream unlimited music to your iPhone for a monthly subscription fee. It's ideal for music fans

For you
Based on a few simple questions, the service will present a selection of new music based on your likes and dislikes

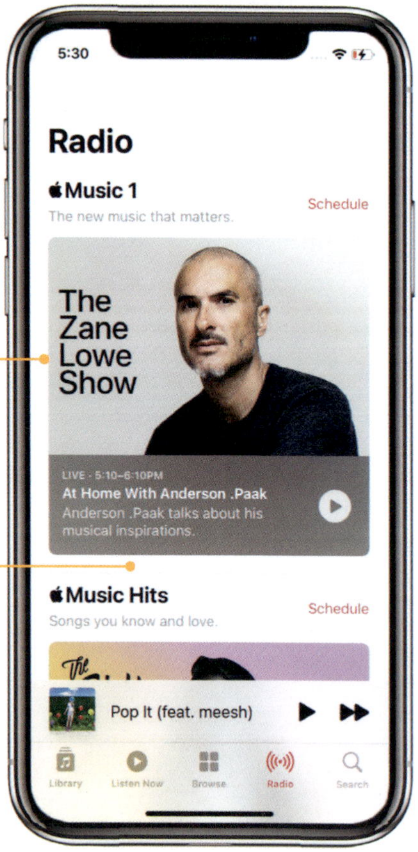

"You can also use the new-look Music app to see what the current hot music is"

Apple Music Sign up to Apple Music

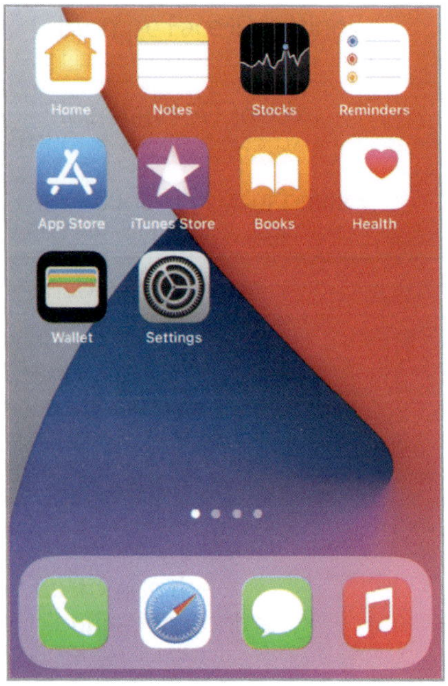

01 Navigate to the app
If you're opening Music for the first time, then you'll find it, by default, in your dock. Once you're up and running with the app, then you'll also be able to access it from the Lock screen and Control Centre.

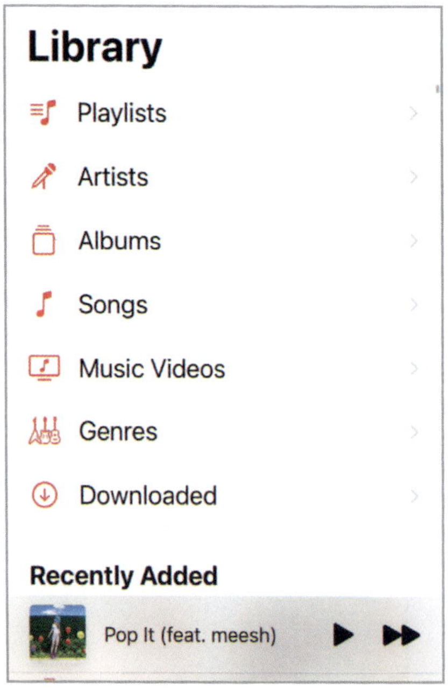

02 Your music
You will notice that your music library and playlists have been condensed into one section called Library. You can access your music or your playlists using the tabs at the top of the screen.

03 Join Apple Music
Tap on the 'Listen Now' icon at the bottom of the interface and you will be prompted to begin your free Apple Music three-month trial. Tap on the prompt to get going.

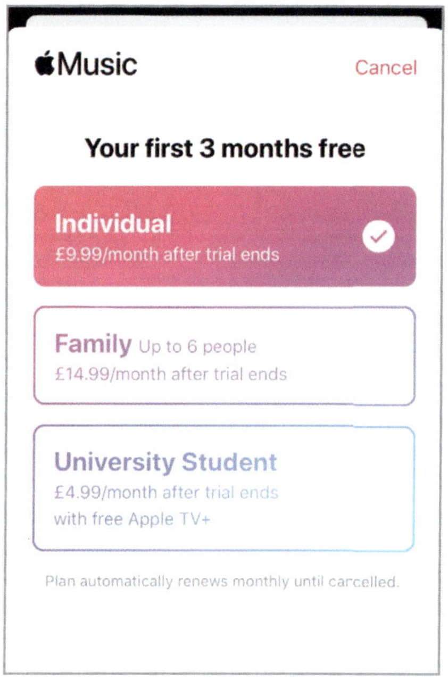

04 Choose a plan

You will then be asked to choose a membership plan, so select either the Individual or Family package option and then enter your Apple ID to complete the process.

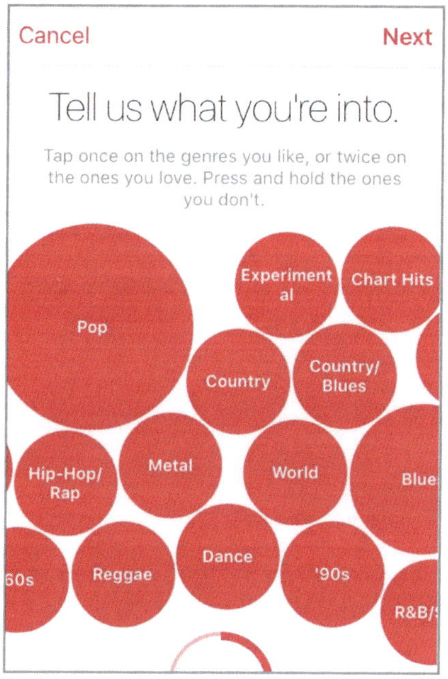

05 Tell the app what you like

On the following screen, tap once on the genres of music that you like and twice on the ones that you love and then press and hold on the genres that you are not a fan of. When done, tap Next.

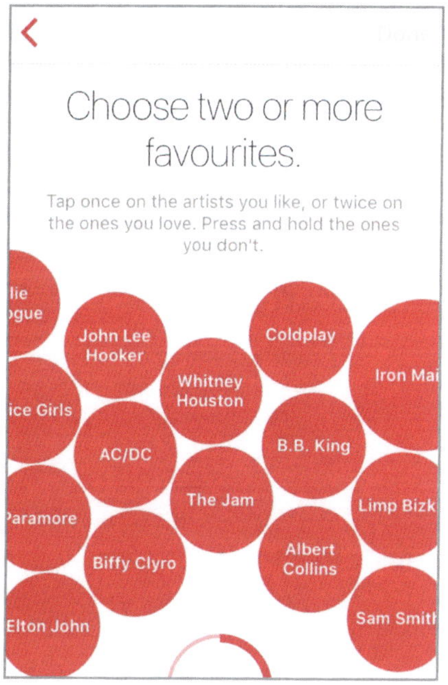

06 Choose your favourites

Next, repeat the process but this time on the selection of bands and artists that the service selects for you based on your previously stated likes and dislikes. Tap Done when you are ready.

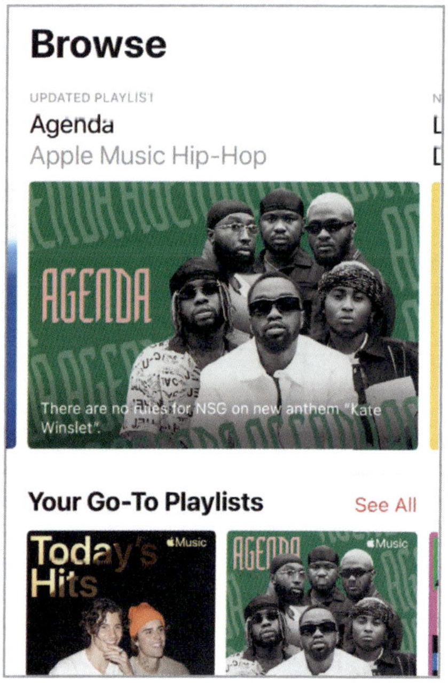

07 Get listening

Based on your likes and dislikes, the Apple Music service will present new music for you to check out and listen to under the Listen Now and Browse sections. Tap on the Play button to start listening.

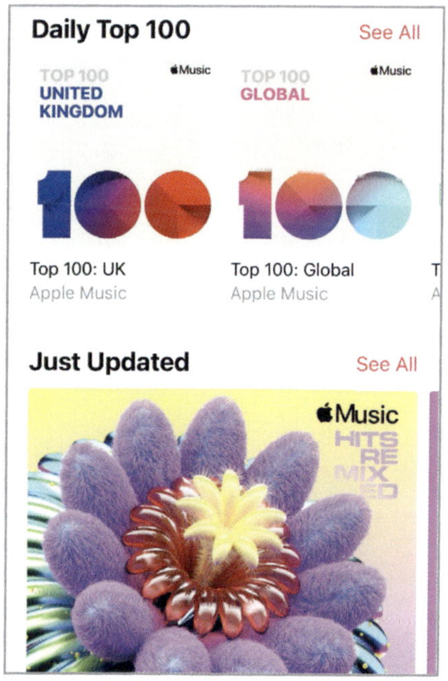

08 Check what's new

Tap on the Browse section to also find new and recently added music. Here you can discover hot tracks and scroll down to check out various playlists. You can also search by keyword.

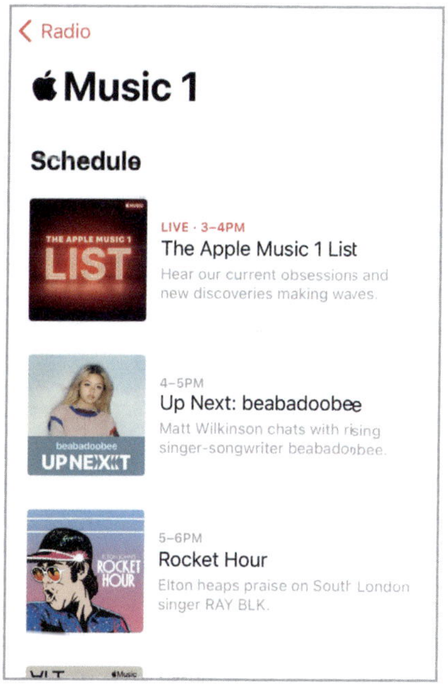

09 Listen to the radio

Apple has three 24/7 radio stations and they are completely free – you don't need to be subscribed to Apple music to listen to it. Tap on the Radio tab to find it and many other stations.

Enhance your audio entertainment with Podcasts

With such a diverse range of shows available, podcasts are sure to become even more popular

The official Podcasts app from Apple is designed to streamline every part of the process of subscribing, downloading and listening to podcasts. Excellent discovery options are included that will offer more than enough choice to get you started. You should bear in mind the potential costs of downloading episodes, particularly video podcasts, but the actual content is free. We will guide you through the options and also how to build up a library of entertainment that will serve you well over time. Podcasts are excellent value, easy to use and Apple has included everything you need to enjoy them on your iPhone.

Discover podcasts
Apple offers a featured listing and also provides top charts for the most popular podcasts, which will help you find new shows to listen to

Your selection
Download any podcast episode you fancy and subscribe to shows. New episodes will download automatically in the background when they are available

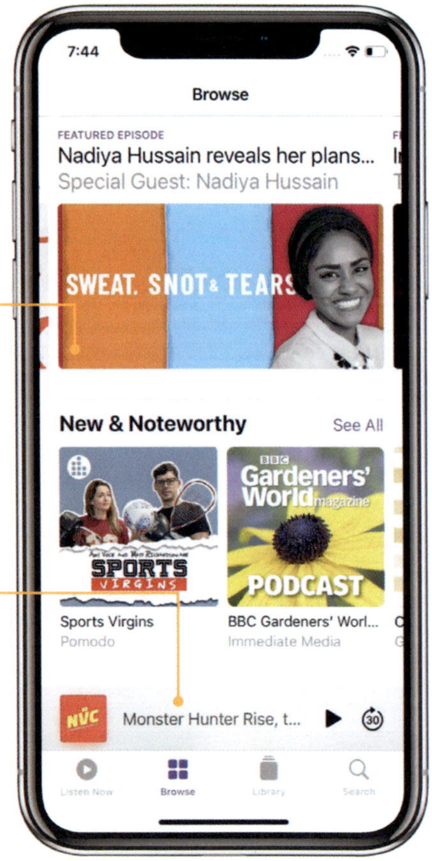

"Podcasts are excellent value, entertaining and easy to use. Apple has included everything you need"

Podcasts Make the Podcasts app work for you

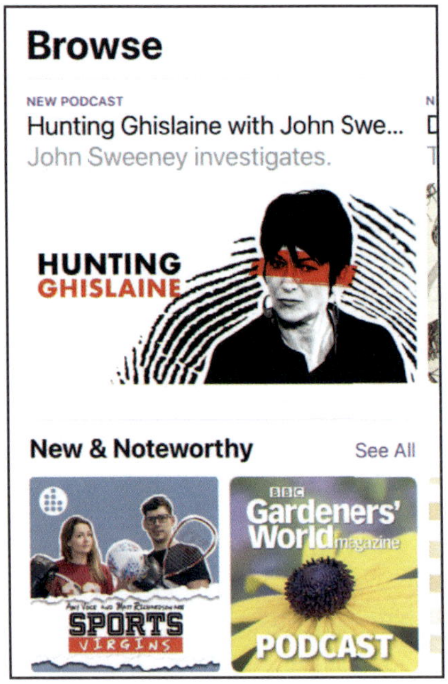

01 Find the app
Open the Podcasts app and then tap the Browse tab. You will now see a selection of popular podcasts – tap one to see more detailed information about it.

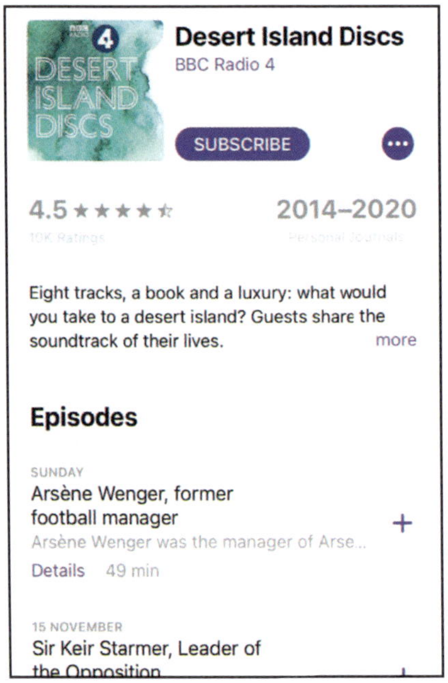

02 The episode listing
The main screen for any podcast should show all available episodes, along with download links. The length of each episode is displayed along with the date it was aired. Now scroll down to the reviews.

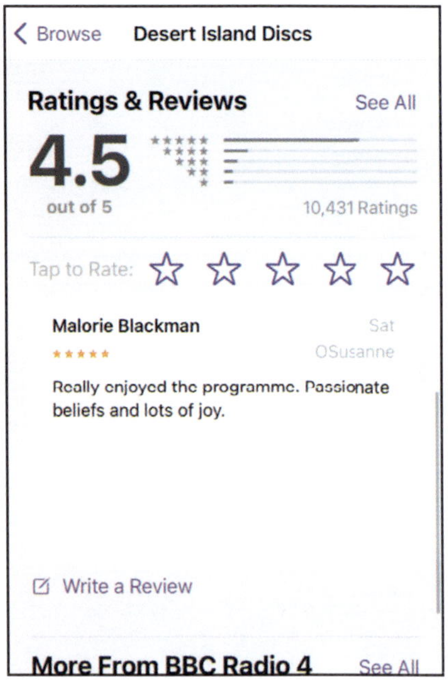

03 Is it worth subscribing?
The Reviews section details the thoughts of others and is a very good indicator as to whether it is worth your time. The more positive reviews, the better, and many are constructive which helps a lot.

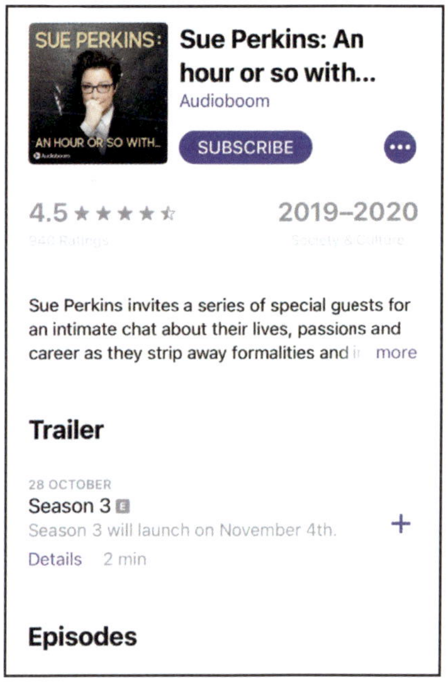

04 Time to subscribe

Tap the 'Subscribe' button to add the podcast to your list. A pop-up will appear to confirm you are subscribed and the latest episode of that podcast will automatically download to your device.

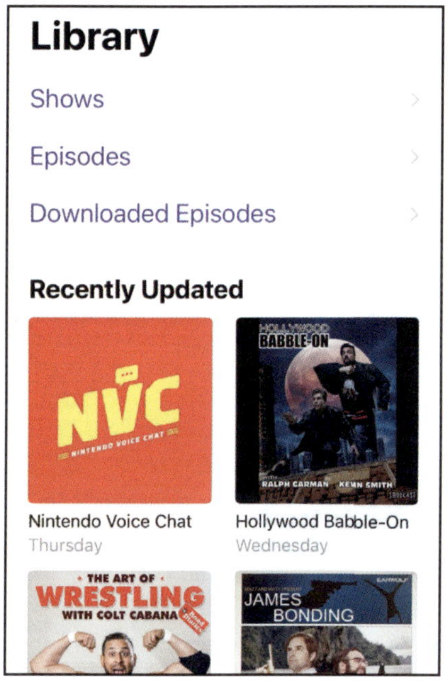

05 Your subscriptions

In the Library section, all of the podcasts you are subscribed to are displayed as thumbnails for easy navigation. Tap on a podcast, then on the more icon if you want to cancel your subscription.

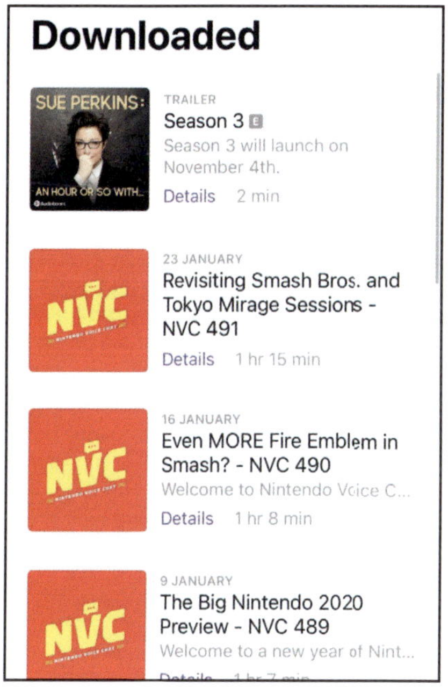

06 Download a podcast

Tap the cloud icon next to an episode to download it. Go to Library> Downloaded Episodes to see how much longer is left. When it's done, you can listen to the show without an internet connection.

07 Time to listen

There are some useful controls on the playback screen; the Speed option lets you adjust the playback speed and you can also rewind by 15 seconds and skip forward 30 seconds.

08 Write a review

In the podcast listing, scroll down, go to Reviews and tap 'Write a Review'. Input a comment and star rating, but try to be as accurate as possible to help other potential listeners.

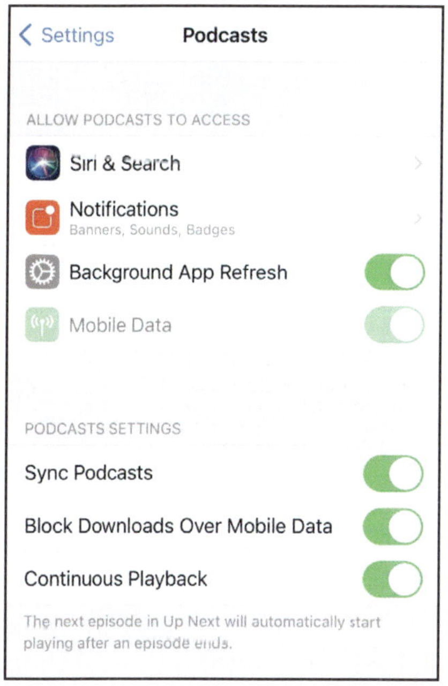

09 Advanced podcast management

Go to Settings>Podcasts where you will see options to customise the podcasting experience. You can download over Wi-Fi only and also limit the number of episodes downloaded automatically.

Use Apple TV

The iPhone is perfectly designed for the mobile movie experience thanks to its high-resolution screen, long battery life and excellent portability

The iPhone's Apple TV app alone has the potential to keep you occupied on long plane journeys, while you are bored in a hotel or waiting for an appointment. It adds an invaluable function to the iPhone experience that could almost justify half its cost straight away. It has been designed for ease of use, as most Apple software has been, and takes great care of many of the niggles that are found in competing devices.

For example, it will automatically start playing a film or television programme from the same point that you last left it, which is handy, and expanding the screen only requires a simple double tap. In true Apple style, everything is designed to help you get the most from your viewing experience. In the age of digital downloads and mobile experiences, Apple really are ahead of the vastly steep upward curve.

In this tutorial, we are going to show you how to obtain new movies and television programmes, download them to your iPhone and make the most of the viewing experience. Undoubtedly, you could easily figure out all of this yourself, but a little knowledge goes a long way and missing out on the movie capabilities of the iPhone would be a real shame, given the benefits it offers.

So, grab your iPhone, put your feet up and check out the following steps.

"Obtain new movies and television programmes, transfer them to your iPhone and make the most of the viewing experience"

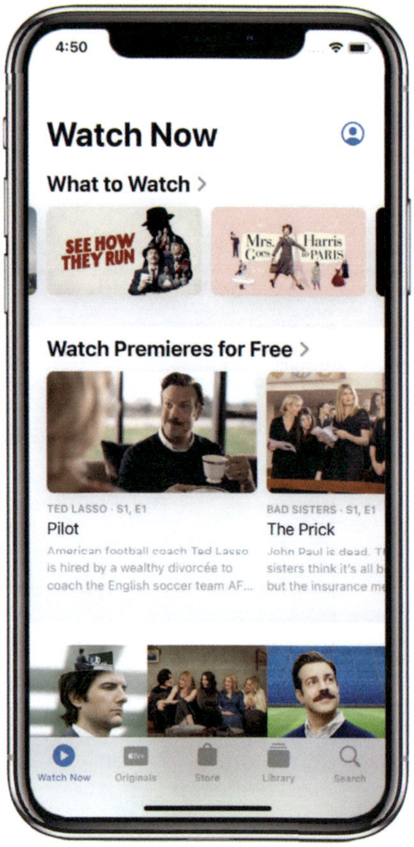

Apple TV Watch movies and TV shows

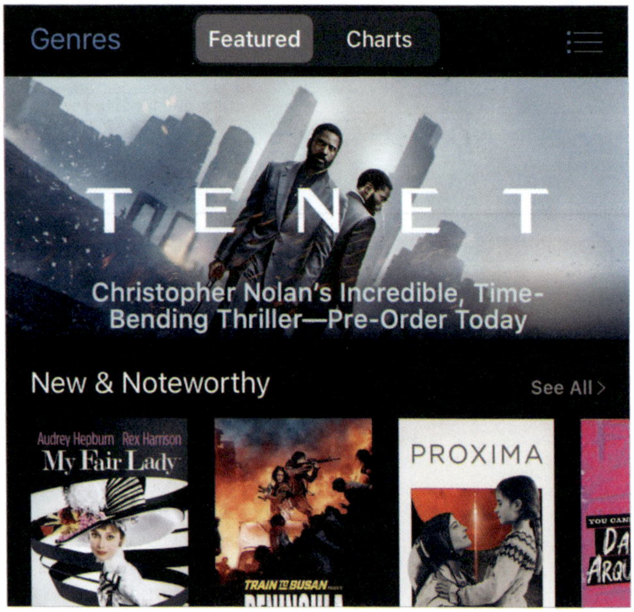

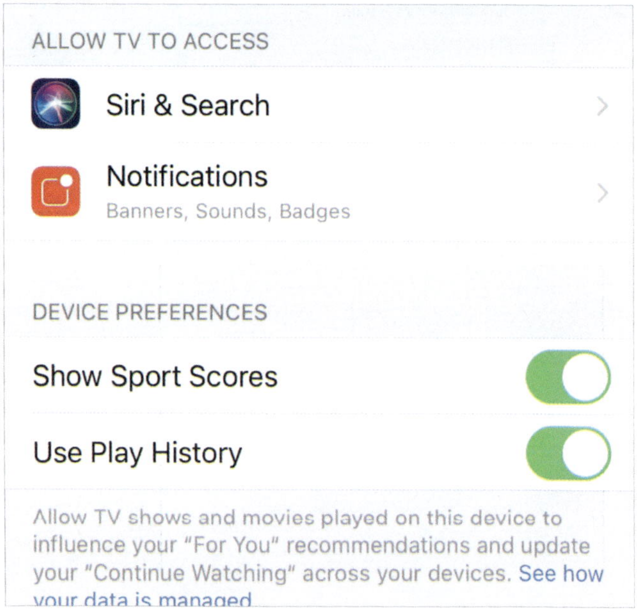

01 Grab a film or TV show

The easiest way to obtain good quality content is via iTunes. Navigate to the Movies or TV Shows section and choose the film or programme you would like to rent or buy. You can also watch trailers for free to help you determine whether the content is worth your money.

02 Get it on the iPhone

Anything you buy on your device will be downloaded and made accessible through your Apple TV app. If you are logged into the same Apple ID on other iOS or macOS devices then all of your purchases will be pushed to those as well thanks to your iCloud Drive.

Watch movies and TV on your iPhone

Get the most out of your portable video

Back where you left off
The iPhone automatically remembers where you finished watching the video and will start any film at that exact place when you open up the app again

Full control
You can move to specific parts of a video by moving the slider with your finger; the further up the screen your finger is, the more precise the movement is

Main controls
The main control keys are standard and are brought up by tapping the screen once. You can play, pause, forward or rewind when you feel the need to

Full screen
If you want to clear the on-screen furniture out of the way to watch in full screen, simply give the main viewing area a tap and it all slides away from sight

High definition
Many iTunes movies and TV programmes are in HD, which offers a crisper view. Sometimes you'll pay more for the video, but think of it like paying more for Blu-ray. However, don't expect to get as much out of HD movies from your iPhone screen as you would your computer monitor or dedicated HD screen.

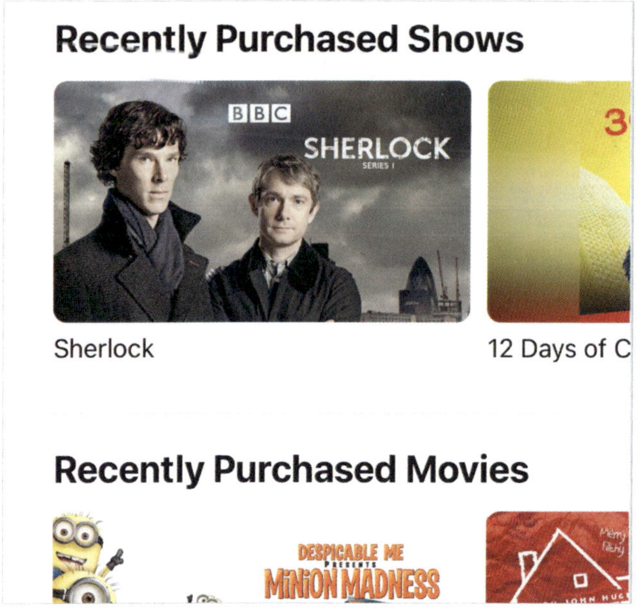

03 Open the Apple TV app

Navigating the Apple TV app is simple thanks to three icons at the bottom of the interface. Tap on Library to find your purchased Movies and TV Shows. If your purchased content isn't on your iPhone, tap the cloud icon next to it to re-download and start watching on your device.

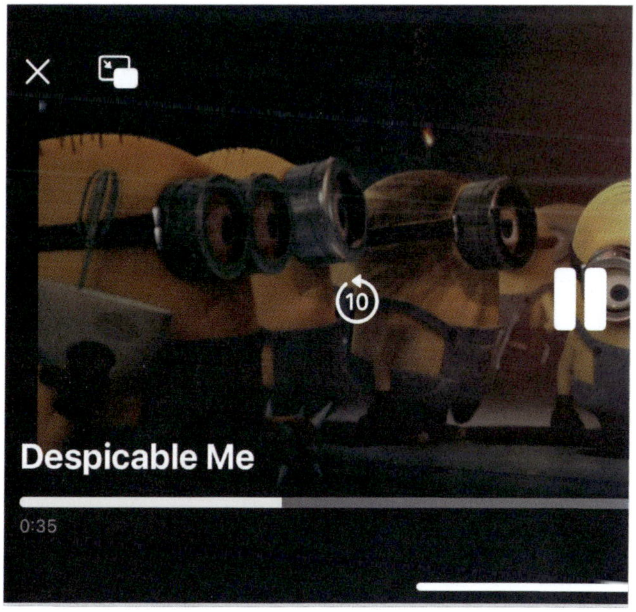

04 Small changes

Double-tapping the screen will make the movie play in full-screen mode, and doing so again will take it back to standard format. The rest of the on-screen actions are obvious in their implementation, such as play, pause, etc. Use the slider at the bottom to navigate to a desired point in the video.

Purchase a book

Learn how to open up to a whole world of digital reading

It is rather easy to get lost in a world of crazy apps and social media, but the iPhone is also a fantastic ebook reader. It has a number of advantages over the Kindle in that it can do a great deal more than a dedicated device, but on a purely ebook-reading scale, the iPhone is still one of the most advanced out there, especially considering its size.

What's more is that Apple already has a tried and tested way of delivering ebooks directly to its device: namely through the iTunes Store. Apple hasn't just bundled the new books into that system, though, because it's created a separate space for these so that users can be absolutely sure of what they are downloading.

The Books app holds all of your ebooks, and from there you can access the custom-built Books Store to make purchases, which get downloaded directly onto your iPhone. What's more, since the iOS 8 update, the app is included as standard. The system is magnificent in its simplicity and, just like the App Store, it makes impulse buys a regular occurrence.

This tutorial will take you through your first download from the Books app so that you can get a feel for the system. It's then up to you to resist buying a library's worth of content on each visit.

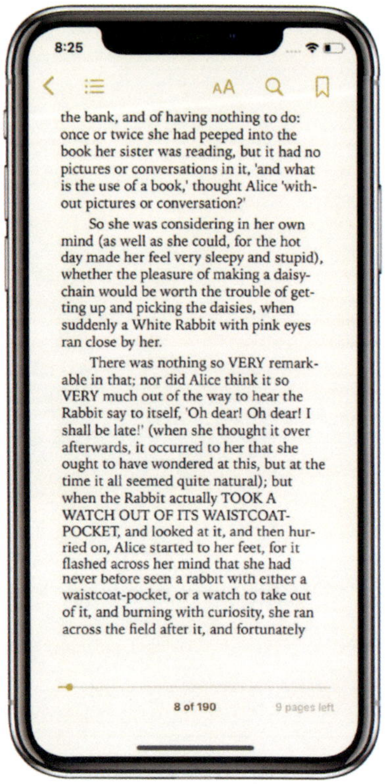

"The iPhone is one of the most advanced ebook readers out there, especially for its size"

Books Purchase a book

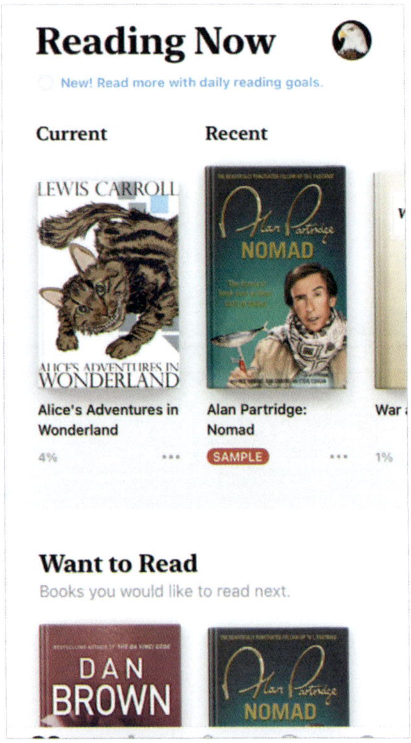

01 Launch the app

The Books app is available on your iPhone by default. If you have purchased a book before, you'll see your most recent book in the Readng Now section. To purchase a book, hit the Book Store tab.

02 Familiar feel

The Books Store is very much like the Podcast Store. Books are categorised and searchable, and everything is charted so that you can see what is selling best.

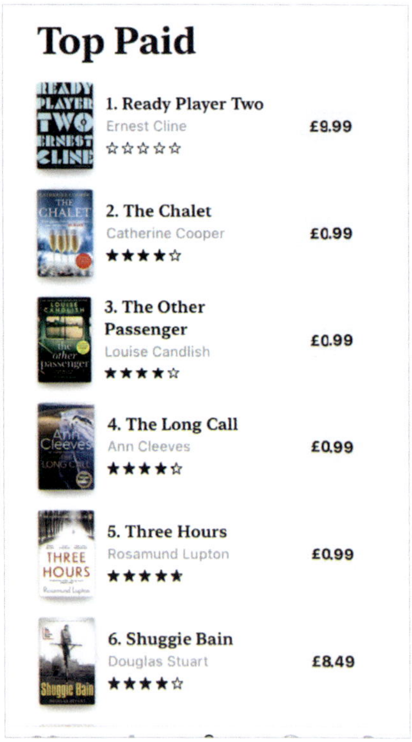

03 Charts

There is even a section on the Book Store where you can see the charts so you are able to pick and download titles from it. You'll find separate lists for free and paid content.

The Books Store homepage

Find your way around your new home for digital books

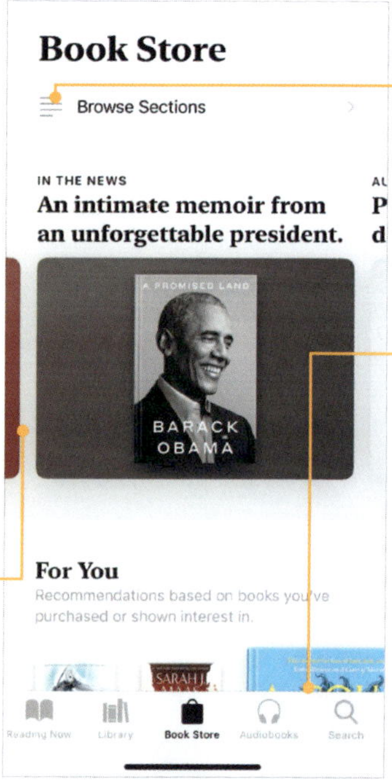

Choose your category
Select on Browse Sections to choose a specific book genre to browse from. This store covers everything from Business and Biography to Reference and Health

Tabs at the bottom
At the bottom of the interface you will find five tabs to help you to navigate through the Store, see what you have already bought, and also what you've recently read.

Easy navigation
Navigating through the Store with your fingers is easy. Swipe left to see more and tap individual books to get more information

Continue shopping
One of the best new features of the iBooks app is that once you have purchased a book, the app will no longer go straight to your library as you did before, leaving you free to continue shopping.

04 Free books

Like the App Store, there are a huge number of free books. These tend to be classics, so you can go ahead and get great content for nothing. There's a ton of free material.

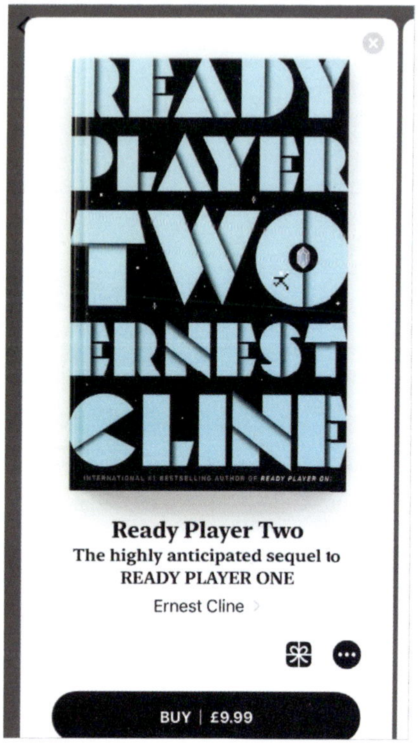

05 Get a book

When you have found a book that you want, tap the price followed by 'Purchase'. You'll be prompted for your password and, after entering it, the book will download.

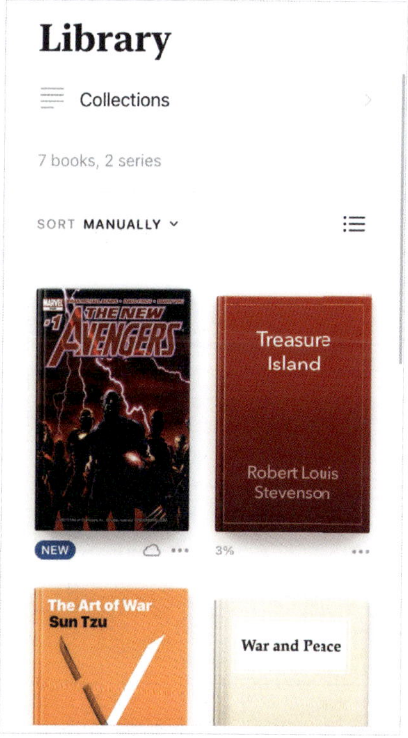

06 New books

Your new books will now appear in the Library section of the app. A circle will begin to fill as it downloads. Once downloaded, the book will become available to read at your leisure.

Find your way around Books

Having an ebook reader on the iPhone is very cool, so here's how to customise it to your liking

When you're reading an ebook on the iPhone it feels natural, and we are certain that we'll be doing a lot more reading now that it's so simple to carry books around with us. The beauty of the iPhone interface means that making changes to the way the Books app looks and feels is very simple.

Users can opt to make text bigger, change the font, change scrolling options, alter the brightness and change the background theme of the book without having to leave the page they're reading. Try performing the same operations on a paper book and you'll see the advantages of reading in a digital format. Follow our quick tutorial on how to get more from the already excellent reading experience.

"Try doing these operations on a paper book and you'll see the advantages of reading in a digital format"

True text
Once you have your fonts and the brightness set up how you like, you can read for hours on this device

Scrubber
You can also navigate through the app using the scrubber bar at the bottom of the page. Just drag your finger along the line

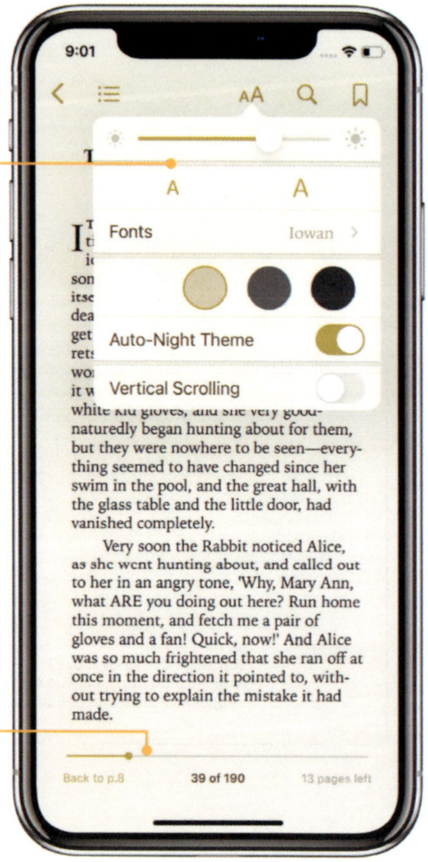

Books Font, size and brightness

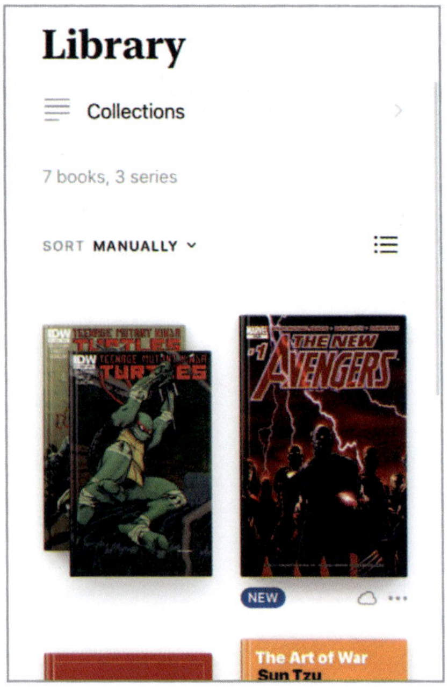

01 Open and browse

Open the Books app, go to Library and then tap on the cover of a book that you would like to read. Then all you need to do is tap the page to toggle the option buttons.

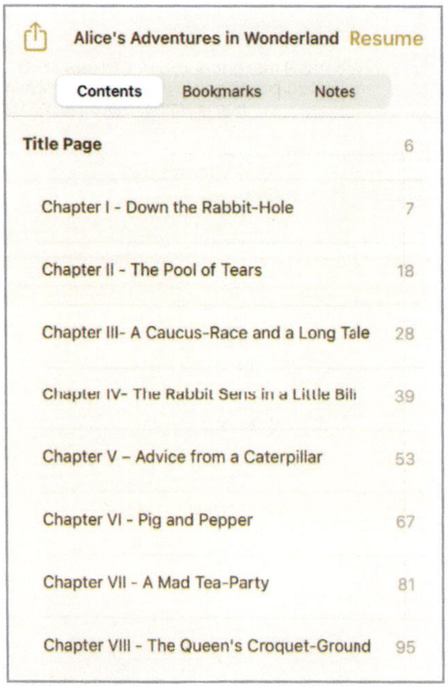

02 Contents

You are able to easily navigate from the contents page to a chapter simply by tapping on it. You can then return to this screen from the second left icon at the top.

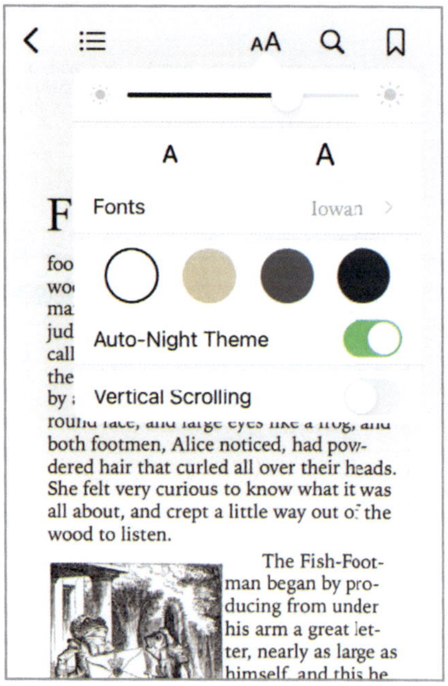

03 Font type

Tap on the 'AA' icon to bring up the options for changing the brightness, font size, fonts and themes, which can be very useful. Tap the Fonts button and pick from the options.

04 Select font
Select a font that you like and find comfortable to read in. Take some time to figure out which font works best for your browsing habits, and then press on it.

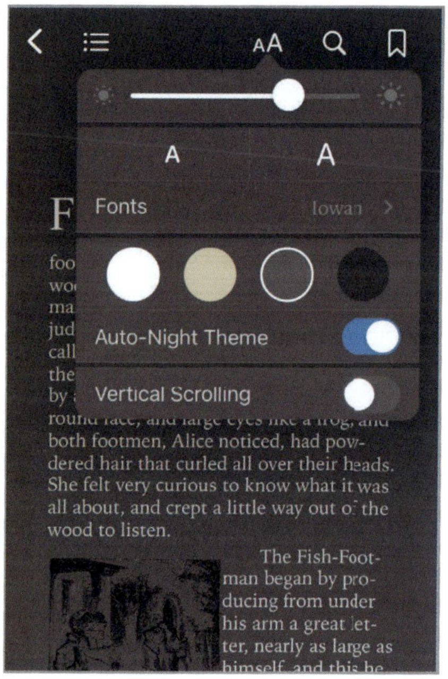

05 Themes
Press on the Theme option in order to pick a style that you are comfortable with. Books allows you to virtually flip pages as if you are reading a regular book, or use endless scrolling.

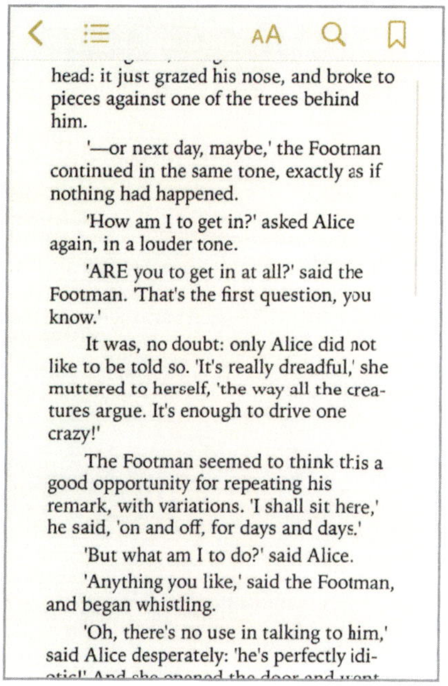

06 Vertical scrolling
Some people are already familiar with reading digital content by swiping up the screen, and the Vertical Scrolling option allows you to do just that, which is great.

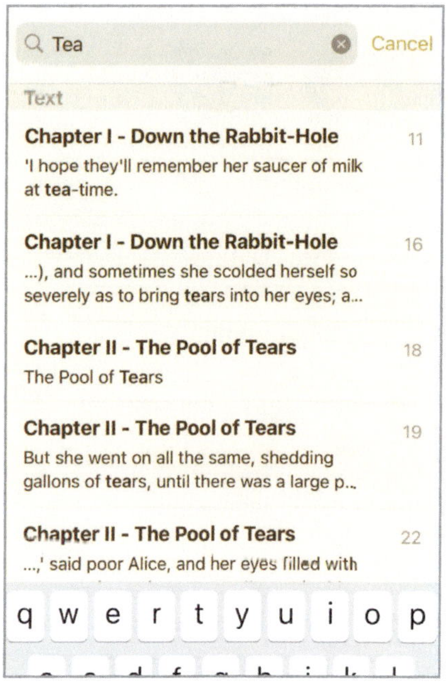

07 Spotlight index
Tap on the magnifying glass in order to bring up a search field. Every book on the store is fully indexed so you are able to instantly find individual words in a book.

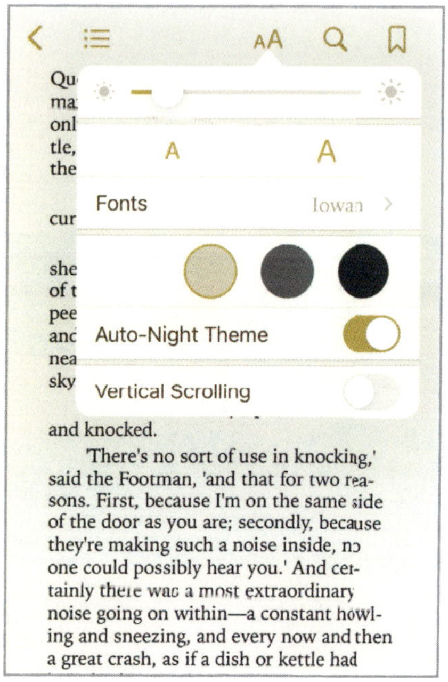

08 Brightness
Tap on the brightness slider in order to adjust the brightness and reading mode. White or Sepia is a good choice for daytime reading and Night for night-time reading.

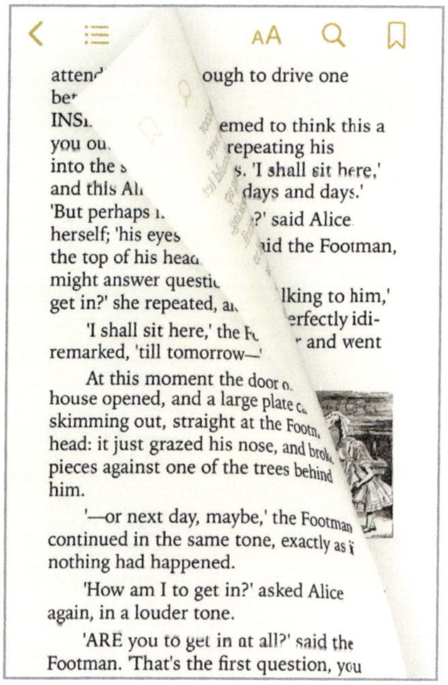

09 Suitable setting
Changing the brightness means that you are able to choose whatever is most comfortable for you and your eyes. Select your favourite settings and they will be retained.

Browse the Files app

Access various cloud storage services through a single app, so you can find all of your documents and images in one place

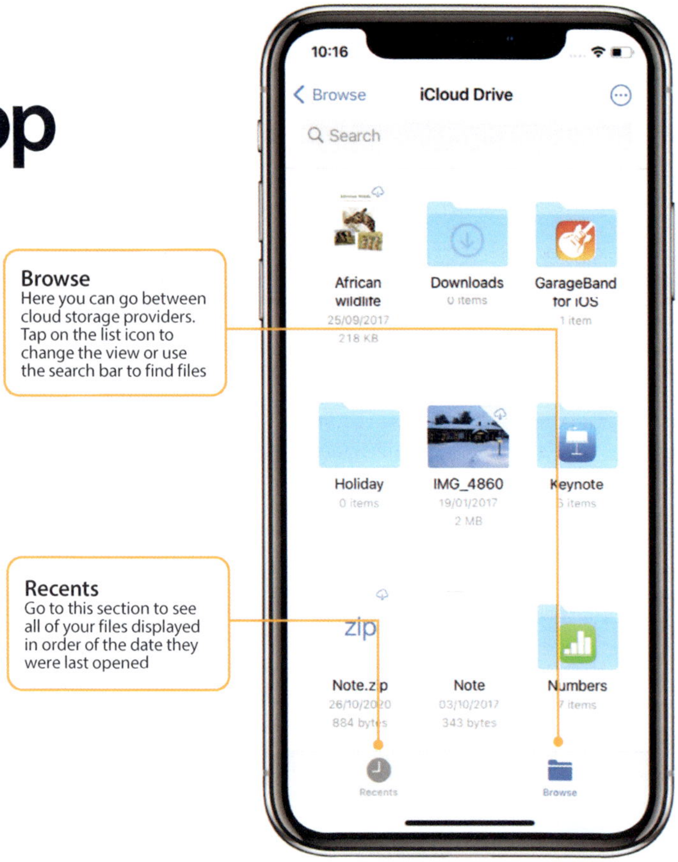

In previous versions of iOS there was the iCloud Drive app; this let you access the files saved to your iCloud account on any Apple device. The Files app is the evolution of this concept. Various cloud storage services such as Dropbox, Google Drive, Microsoft OneDrive and Apple's own iCloud Drive can all be accessed through Files. This means that, for example, you can save a Word document to Dropbox on your PC and then pick it up on your iPhone – through files – while on the go. Documents are also made accessible outside the confines of the app that created them. For example, a file created in the Notes app can be opened in Pages. There's also a range of tools that makes arranging all of your documents while on-the-go much easier than it has been in the past. It's never been more convenient to access and organise your content stored in the cloud.

> "It's never been easier to access and organise your content"

Files Learn how to use the Files app

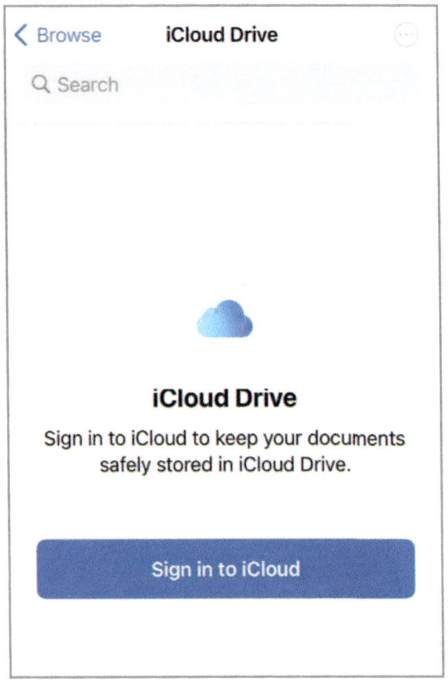

01 Get started
If you aren't already signed into iCloud, you will be prompted to when you boot up Files for the first time. Tap on 'Sign in to iCloud' and you will be taken to the login screen.

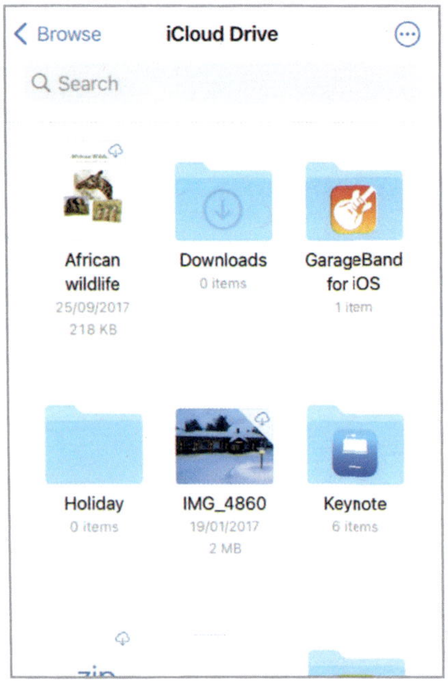

02 Browse Files
Now you can look through all the files you have saved in your iCloud Drive. If you have any of the iWork apps (such as Pages) on your iPhone, these files can also be found here.

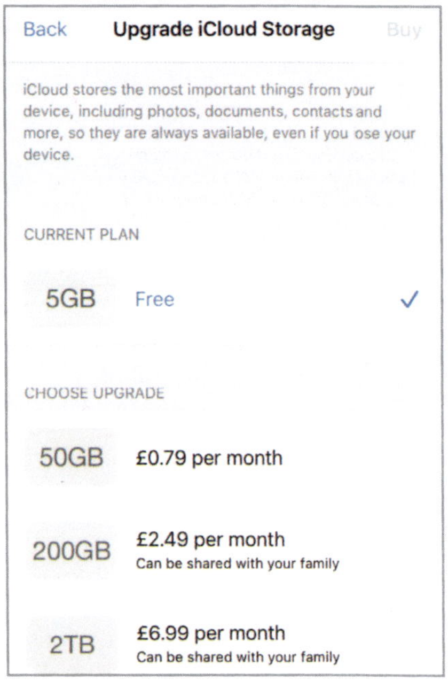

03 Upgrade iCloud storage
As you fill iCloud Drive with files, you may run out of space. Go to Settings>Apple ID>iCloud>Manage Storage and tap on Upgrade. You can pay a monthly fee for 50GB, 200GB or 2TB.

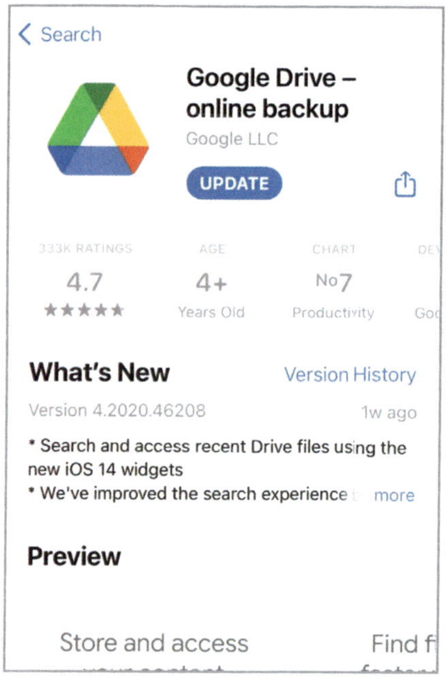

04 Find Apps

To add cloud storage providers to Files, go to the App Store and search for apps. Dropbox, Microsoft OneDrive, Google Drive and more can all be downloaded here.

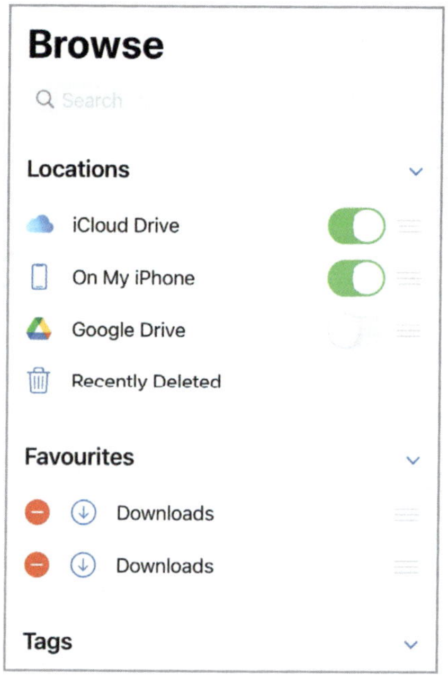

05 Add to Files

Once you have downloaded the apps, go to the Browse section of Files, tap the more icon at the top of the screen and then Edit. Move the sliders to the right to allows that cloud storage to be accessed.

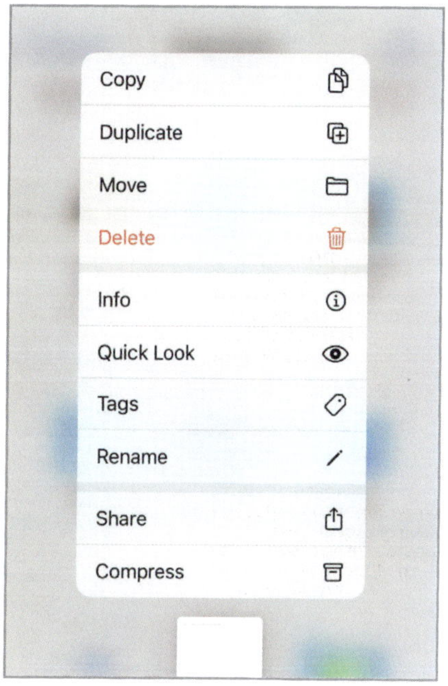

06 Tag files

Long-press on an item to bring up a menu, tap on the right arrow to find the Tags option. You can create a new tag or use an existing one. Tap on the Edit option in Browse to delete or rename tags.

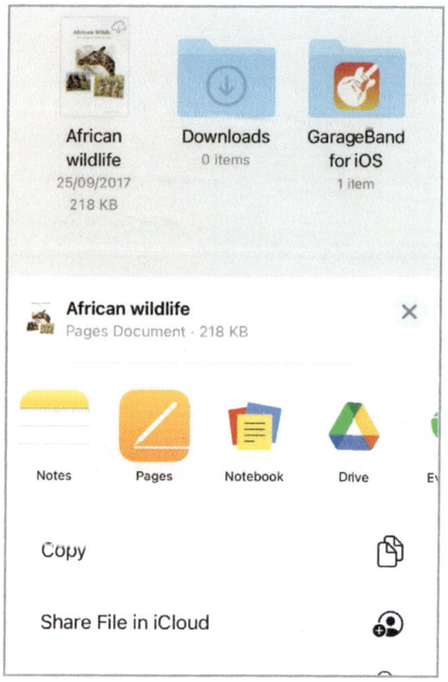

07 Open in a different app

To move a document to another app, long-press on the file, select Share and then choose the app. In this example we are moving a document from Pages to Notes by tapping on the Notes option.

08 Edit images

Tap on the pen icon in the top-right corner when viewing images to use Files' drawing tools. There are three drawing tools to choose from or tap the Plus icon to add text or other a signature.

09 Make Changes

Any changes you make to a document or an image will be automatically saved in the cloud, so you can close the Files app and immediately open the file on another device.

Set up your Apple Watch

The iPhone is an essential tool when setting up your new Apple Watch from scratch

The Apple Watch is an exciting piece of kit, especially if it's your first smartwatch purchase. While many users will want to delve into its suite of features immediately, first you'll need to use your iPhone to correctly set it up. The two devices work in tandem, but due to the various menus and options on the iPhone, it's an essential piece of kit to have to hand when you first venture on to your Apple Watch.

Although the setting up process is easy enough, the time it takes is a whole different story. Due to the intricate components of the smartwatch, and the technology it takes to both sync and pair with your iPhone, the process takes around 30 minutes. However, this time can vary depending on some of the options available during the process.

The various parts of the set up process is aimed at making sure that both your iPhone and Apple Watch work in perfect harmony. While understanding the various menus on offer is simple enough, it's paramount that you carefully select the options you need through the Apple Watch app on iPhone. The last thing you want is to have your Watch set up entirely in Spanish, with the time zone set to Mongolia. The time you spend ensuring that you've correctly gone through on each menu item will save you so much hassle later down the road, so it's well worth the effort.

Once you've got the Watch up and running, don't think the set up process is complete. Instead there are some initial settings you'll want to customise to make sure your Watch is running at optimum speed. For this you'll be relying on the iPhone's Apple Watch app, which includes various menus that enable users to tinker with the layout of their watch, manage its brightness and other key settings.

But don't fret if this sounds like a lot of work, as we've laid out a quick setting up guide to outline all the steps you need to take to get your Apple Watch set up and working with your iPhone.

"The set up process is aimed at making sure that your iPhone and Apple Watch are working in perfect harmony"

Configure your Apple Watch

Make use of the amazing features in the iPhone Apple Watch app to get more from your new wristwear

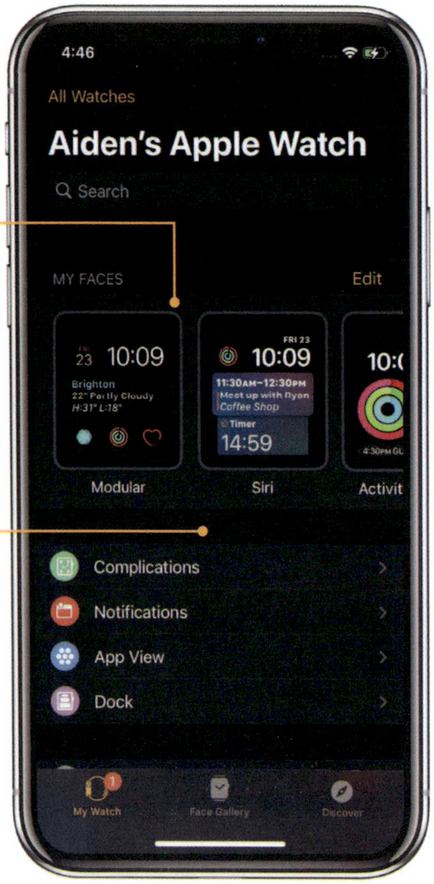

Alter settings

One of the great things about the iPhone is that you're able to alter all Apple Watch settings from within the official Apple Watch iPhone app

Start customising

Another big addition to the Apple Watch app on iPhone is being able to customise the app layout screen. You can move the icons into different positions and alter locations on the launcher

Default look

This is the default look of the Apple Watch when you turn it on. You can customise it either on the display itself, or through the iPhone Watch app

Settings Use your iPhone to set up Apple Watch

01 Start pairing
When you first turn on the Apple Watch, you'll be prompted to select the default language. After that, you'll be taken to the pairing screen, where you'll need to start connecting your iPhone to the watch.

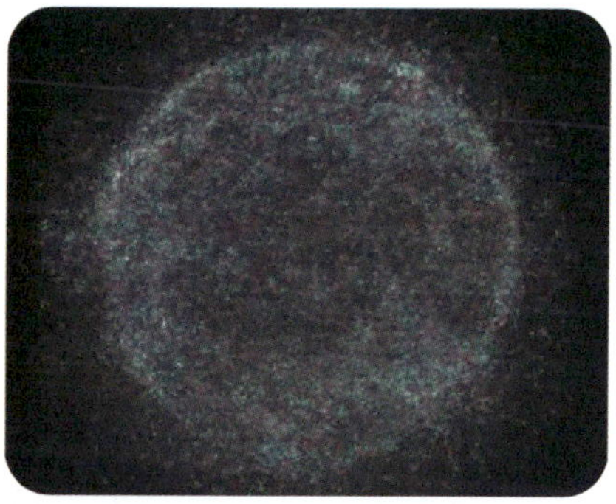

02 Find in Viewfinder
To confirm the pairing process, you'll need to hold your iPhone up directly in front of the Apple Watch. Keep your iPhone as still as possible, as it can be a bit tricky to line it up correctly.

03 Start to sync
Your Apple Watch will now begin to fully sync with your device. The process can take as long as 20 minutes to complete. During this time you can explore some of the extra options on your iPhone.

04 Choose wrist preference
After syncing, the first screen that will greet you on your iPhone will ask you to choose your wrist preference. No matter what you choose, you can change this at a later point if necessary.

05 Login to iCloud
The next screen will prompt you to login to your iCloud account. Make sure to use the correct login details that you created when you first set up your iPhone so that the two devices pair correctly.

06 Read tutorials
Over the next series of iPhone screens, you'll need to read about the various Apple Watch features. It's important to read these as it'll teach you about both app and Siri integration.

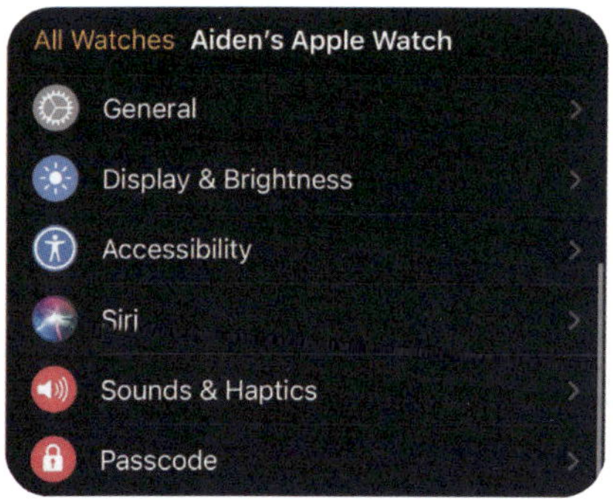

07 Watch overview
After pressing the confirm button, you'll now be taken to the home screen of the Apple Watch app for iPhone. There are various menus you can explore to tinker with specific Apple Watch settings.

08 Configure Home screen
Your Apple Watch should now show the default home screen that has been set. Long press on the display, known as a Force Touch, to bring up other home screen designs that you can choose to use.

09 Move icons
With a single press on the upper side button, known as the Digital Crown, you'll be taken to your watch's app launcher. Force Touch on any icon to relocate it to a different area of the launcher.

✓ Get great savings when you buy direct from us

✓ 1000s of great titles, many not available anywhere else

✓ World-wide delivery and super-safe ordering

Printed in Dunstable, United Kingdom

91222526R00076